Power Electronics and Power Systems

Series Editors
Joe H. Chow, Rensselaer Polytechnic Institute, Troy, New York, USA
Alex M. Stankovic, Tufts University, Medford, Massachusetts, USA
David J. Hill, University of Hong Kong, Pok Fu Lam, Hong Kong

T0171879

The Power Electronics and Power Systems Series encompasses power electronics, electric power restructuring, and holistic coverage of power systems. The Series comprises advanced textbooks, state-of-the-art titles, research monographs, professional books, and reference works related to the areas of electric power transmission and distribution, energy markets and regulation, electronic devices, electric machines and drives, computational techniques, and power converters and inverters. The Series features leading international scholars and researchers within authored books and edited compilations. All titles are peer reviewed prior to publication to ensure the highest quality content. To inquire about contributing to the Power Electronics and Power Systems Series, please contact Dr. Joe Chow, Administrative Dean of the College of Engineering and Professor of Electrical, Computer and Systems Engineering, Rensselaer Polytechnic Institute, Jonsson Engineering Center, Office 7012, 110 8th Street, Troy, NY USA, 518-276-6374, chowj@rpi.edu.

More information about this series at http://www.springer.com/series/6403

Slobodan N. Vukosavic

Grid-Side Converters Control and Design

Interfacing Between the AC Grid and
Renewable Power Sources

 Springer

Slobodan N. Vukosavic
Department of Electrical Engineering
University of Belgrade
Belgrade, Serbia

Serbian Academy of Sciences and Arts
Belgrade, Serbia

ISSN 2196-3185 ISSN 2196-3193 (electronic)
Power Electronics and Power Systems
ISBN 978-3-030-10346-0 ISBN 978-3-319-73278-7 (eBook)
https://doi.org/10.1007/978-3-319-73278-7

Printed on acid-free paper

This Springer imprint is published by the registered company Springer International Publishing AG part of
Springer Nature.
The registered company address is: Gewerbestrasse 11, 6330 Cham, Switzerland

*To my parents, Anka and Nikota Vukosavić,
whose love and guidance remain with me.*

Preface

This textbook is intended for undergraduate students at final years of electrical engineering. It is also recommended for students preparing capstone project where they need to understand, model, supply, control, and specify grid-side converters in micro-grids, distribution grids, and residential, commercial, and industrial applications. At the same time, it can be used as a valuable reference for other engineering disciplines involved with smart grids, renewable sources, and energy accumulation. It is also suggested to postgraduates and engineers aspired to development of electronically controlled production, transmission, distribution, and use of electrical energy. Unlike majority of textbooks on this subject, this book requires a rather limited background. Wherever possible, an effort is made to provide the text approachable to students and engineers in engineering disciplines other than electrical.

The scope of this textbook is to provide basic knowledge and skills in designing and controlling the grid-side inverters, in the electronically controlled static power converters that are becoming a key part in smart grids, as well as in new, power electronics-based systems for production, transmission, distribution, and use of electrical energy. Basic engineering considerations are used to introduce the salient features of grid-side power converters in intuitive manner, easy to recall and repeat. The book prepares the reader to comprehend the key properties and roles of grid-side converters, to analyze their steady-state and transient characteristics, to obtain basic notions on their design and control, to evaluate the effect of electronically controlled loads in ac grids, to understand the changes introduced by replacing traditional synchronous machines by source-side converters, to consider the effects of an increased number of power electronic devices on power system controls and protections, and to foresee the future developments and related requirements.

Discussion, analysis, and knowledge development on grid-side power converters and the introduction of basic skills are suited for the electrical engineering students in their final year of undergraduate studies. Required background includes mathematics, physics, and introductory courses of power electronics, control theory, and power engineering. The textbook is made accessible to readers without the advanced background in the subject matters. An attempt is made to prepare a self-contained book. The book comprises questions and answers and solved problems wherever the learning process requires an overview. The chapters comprise an appropriate set of exercises, problems, and design tasks, arranged to reinstate and use the relevant knowledge. Wherever it is needed, the book includes extended reinstatements and explanations of the required skill and prerequisites. The approach and method used in this textbook comes from the 22 years of author's experience in teaching electrical engineering at the University of Belgrade.

Readership

This book is primarily suited for the electrical engineering undergraduates in their final year. It is also suggested to postgraduates of all engineering disciplines that plan to major in power electronics, renewables, smart grids, and other areas focused on developing a novel, power electronics-based production, transmission, distribution, and use of electrical energy. The book is also recommended to students that prepare capstone project in one of the areas. The book may also serve as a valuable reference for engineers in other engineering disciplines that are involved with the subject matter.

Prerequisites

Required background includes basic mathematics, physics, engineering fundamentals, as well as introductory courses of power electronics, control theory, and power engineering. The textbook is made accessible to readers without the advanced background in these disciplines. It is suited for the students of electrical engineering on their last year of undergraduate studies. Discussion, analysis, and knowledge development on grid-side power converters and the introduction of basic skills are also suited for other engineering students on their postgraduate studies. Individual chapters comprise questions and answers, as well as solved problems wherever it is required by the learning process. Wherever it is needed, the book includes extended reinstatements and explanations of the required skill and prerequisites.

Objectives

- Using basic engineering considerations to introduce principles of energy conversion within basic topologies of grid-side converters
- Providing relevant knowledge and skills in pulse width modulation of two-level and multilevel converters. Comprehending the PWM noise, the current ripple, and the necessary passive filters

- Acquiring skills in analyzing and designing stationary frame, synchronous frame, and resonant current controllers. Comprehending the controller capability to suppress the input disturbances and the voltage disturbances, to provide a well-damped, overshoot-free step response, and to secure the desired robustness
- Studying the process of the feedback acquisition, understanding the basic current sensors, recalling the sampling process and the sampling theorem, designing anti-aliasing filters, and introducing oversampling, decimation, and one-PWM-period averaging
- Mastering the skills of modeling discrete-time current control systems, designing the controller structure, compensating the transport delays, and deriving the optimum feedback gains
- Analyzing the need to suppress the low-order harmonics, comprehending the role of the voltage-disturbance rejection of the current controller, introduction of the active resistance feedback, and designing the controller which encompasses the active resistance feedback
- Studying and apprehending the control of the active and reactive power injected into the ac grid by means of d-axis and q-axis current
- Understanding control of the dc-bus voltage between the two back-to-back inverters used in a wind-power converter set which includes the generator-side inverter and the grid-side converter
- Analyzing the synchronization of the grid-side converter by means of the phase-locked loop. The impact of the phase-locked loop on the power transients caused by the grid frequency change. Control means for the grid-side power converters which emulate the transient response of conventional synchronous machine
- Understanding injection of parasitic dc-bias into ac grids. Study of sensitivity of distribution transformers to the bias. The sensors for detection of small dc-bias voltages embedded into ac voltages. Active suppression of the dc-bias voltages by means of grid-side converters equipped with the dc-bias sensors and the bias-suppression algorithm

Teaching Approach

- The emphasis is given to the system view, explaining external characteristics of grid-side converters. The basic functionality and controls are of the main interest. Design and construction aspects are of secondary importance.
- Where needed, introductory parts of teaching units comprise repetition of the required background which is applied through solved problems.
- Mathematics is reduced to a necessary minimum.
- The main goal is development and use of mathematical models and transfer functions of grid-side converters. At the same time, the focus is kept on physical insight of the power conversion processes.

- Although the hardware design is out of the scope, some most relevant concepts and design skills are introduced and explained. The book also explains some secondary effects, indicating the cases and conditions where the secondary phenomena cannot be neglected.

Field of Application

This book discusses power electronics, power systems, smart grids, and other application of electronically controlled static power converters in production, transmission, and distribution of electrical energy. Analysis and considerations are focused on basic functionality, topologies, controls, and static and dynamic characteristics of power electronic systems used in conjunction with micro-grids, smart grids, renewable sources, static transformers, bus converters, load-side converters, accumulation devices, static VAR compensators, and other similar applications. Contemporary trends in power systems include the introduction of dc voltages and currents in transmission and distribution of electrical energy, the integration of advanced communication networks that enable remote commands and controls, distributed generation of electrical energy, distributed accumulation resources, and new, online negotiated energy transactions that contribute to the benefit of both the utilities and the clients.

Design, control, and use of power electronic systems in electrical power applications requires the knowledge and skills in power converter topologies, pulse width modulation algorithms, acquisition and filtering of the feedback signals, designing and tuning of the digital current controller, controlling the active and reactive power, suppressing the low-frequency line harmonics, measuring and suppressing the parasitic dc-bias in ac grids, using the phase-locked loop to synchronize the grid-side converters to the ac grid voltages, and setting up the converter controls so as to mimic the behavior of conventional synchronous machines. This book comprises considerations, analysis, studies, and examples that facilitate the engineering efforts in designated area.

Belgrade, Serbia   Slobodan N. Vukosavic

Acknowledgment

The author is indebted to Prof. Aleksandar Stanković, Dr. Igor Cvetković, Ing. Ljiljana Perić, Dr. Petar Matić, Dr. Žarko Janda, Ing. Luka Lopin and Dr. Željko Despotović who read through the first edition of the book and made suggestion for improvements.

The author is grateful to his young colleagues, teaching assistants, postgraduate students, PhD students, and young professors who provided comments and suggestions, read through the chapters and commented the questions, problems, solutions and index terms.

The author would also like to thank Darko Marčetić, Nikola Perić, Branko Blanuša, Petar Grbović, Dragan Mihić, and Nikola Lepojević.

Contents

List of Figures

Chapter 1
Introduction

The power systems, power engineering, and power industry are all changing in rapid manner. There are several contemporary drives for such a change. The exhaustion of fossil fuels, the introduction of renewable sources, the increase in electrical vehicles, distributed generation, distributed accumulation, real-time negotiated energy trans-actions, and the introduction of electronic controlled loads and sources with both local and remote supervision and control are all calling for fast and substantial changes in generation, transmission, distribution, and use of the electrical energy.

One of the consequences is the appearance of grid-connected power converters. Most grid-connected power converters that are used next to the electrical loads are the three-phase devices that make use of semiconductor power switches and the pulse width modulation in order to turn the energy of line-frequency ac currents into the energy of dc currents or the energy of ac currents with different frequency. A number of renewable sources also use the grid-connected power converters, and these are the three-phase devices that turn the energy of the primary source into the energy of line-frequency ac currents. Grid-side power converters that serve as the power interface toward the battery-based accumulation units are capable of converting the energy in both directions. In addition to load-side and source-side power converters, there are also the bus-side power converters. Their role is the energy conversion of electrical energy between the two bus systems, whether the ones with different voltage levels or also different types of voltages (ac and dc).

An increased number of source-side, load-side, and bus-side converters change some fundamental characteristics of ac grids. In places where traditional grids included the power transformers, synchronous generators, and capacitor banks, there are an ever-growing number of grid-side power converters, static VAR compensators, and other devices that employ power electronics. Behavior of the grid-side power converters in steady-state and transient conditions is not the same as the one in the traditional equipment. The power sharing capability, the voltage control, the frequency control, the short-circuit behavior, and the other important features of electronically controlled grid-side converters can be fundamentally different. In order to implement, perform, and maintain the key control function of the grid and

S. N. Vukosavic, *Grid-Side Converters Control and Design*, Power Electronics and Power Systems, https://doi.org/10.1007/978-3-319-73278-7_1

key protection mechanisms of the grid, it is necessary to study the structure, the control loops, and the behavior of the grid-side power converters. This study is the prerequisite for adapting the basic grid controls and protections to an ever-increasing number of grid-connected power converters.

1.1   DC and AC Grids

Development of transmission and distribution grids also includes the cases where the ac grids are replaced by dc grids. Therefore, the question arises whether all the ac grids will be converted into dc grids. If this is true, then there is no need to study the impact of ac-type grid-side power converters on ac grids. Instead, the research and development efforts should be focused on developing dc grids with corresponding power converters.

1.1.1   Conventional AC Grids

At the end of the nineteenth century, the power transmission and distribution adopted the system with ac currents. For the efficient operation of electrical power system, the voltage level has to be changed on the way from the electrical source through the transmission and distribution grids, down to the electrical loads. At that time, the only way of altering the voltage level were power transformers which operate with ac currents. At the end of the twentieth century, new semiconductor technologies allowed for the development and use of power electronic devices and systems capable of stepping-up or stepping-down dc voltages, thus providing the means of designing and using the dc transmission lines, as well as the dc distribution lines. At this point, the question arises about the future development of the grids.

1.1.2   DC Transmission Lines

The use of dc transmission lines started since 1960, primarily for the power supply to remote islands over relatively long underwater cables. The dc transmission line Moscow-Kashira of 30 MW, 200 kV started in 1951, while the Swedish line to the island of Gotland of 20 MW started in 1954. An attempt to use a long underwater ac transmission with high-capacitance coaxial cables results in an excessive reactive power, which makes any underwater ac line over 50 km rather inconvenient. In such cases, transmission is performed by high dc voltages and with high-voltage dc cables with ac/dc converter stations on both sides. DC transmission lines were also used to connect two ac systems with two different frequencies or the two ac systems that are not synchronized.

Another benefit of dc transmission lines is the capability to transmit the power over very long distances. With ac transmission lines, the efficient transmission is feasible along the distances comparable to one quarter of the wavelength of the line-frequency electromagnetic wave. With 50 Hz systems, the distance is roughly 1500 km. Recent progress in renewable power sources calls for the power transmission over much larger distances. Volatile and unpredictable power that comes from the wind power and solar power plants makes it quite difficult to establish the balance between the power of electrical loads and generated power. The power balance can be hardly achieved in small areas. In larger areas, the number of all the renewable sources is larger, and their total power has a smaller relative variance. In a hypothetical case where the super-grid with high dc voltages would connect all the continents, a certain number of solar plants would be exposed to the sunlight at all times, and the total power from all the renewable power sources would be rather constant. In this case, the accumulation capacity required to settle the imbalance between the energy production and the energy consumption would be considerably lower.

1.1.3   DC Distribution

In commercial, residential, and even industrial area, most electrical loads run on dc power. An integral part of the household appliances, computers, video and audio equipment, and even electrical drives is the ac/dc converter, the rectifier which receives the ac power from the grid and provides the dc power.

In electrical drives, 3×400 V line voltages are brought to the six-pulse diode rectifier, which provides the dc voltage in excess of 500 V. The dc voltage is brought to variable frequency inverters with switching power transistors which produce the ac voltages and currents required to run the motor. In electrical drives with regenerative braking, the six-pulse diode rectifier is replaced by another inverter which serves as bidirectional power interface between the ac mains and the internal dc-bus.

Most computers and computer accessories have an external or internal ac adapter. The adapter is usually a switched-mode power converter which receives single-phase line-frequency voltage and provides the dc voltage of 5, 12, and 19 V or some other voltage levels. Most contemporary adapters are of PFC type, namely, they have the power factor correction topology and they draw sinusoidal current from the mains.

At this point, it is justified to consider the use of dc wiring in households, offices, and commercial spaces. Namely, if most of the loads eventually make use of dc voltages, it is reasonable to substitute ac mains with dc voltages, avoiding in this way the ac/dc power converters. For low power loads, a reasonable voltage level of +48 V simplifies electrical insulation, electric shock protection, and short-circuit protection. For large power industrial loads, such as the electrical drives, the dc voltage has to be larger than 300–400 Vdc.

In certain buildings, factories, commercial blocks, and even large ships, the loads are supplied from the small local grid. The local grid is often a dc grid. This grid could include the batteries and other energy accumulation devices and some local energy sources such as the micro-turbines, solar panels, or wind turbines, and it has large three-phase grid-side inverters as an interface with the ac grid, which serves as the main supply. These are the examples of efficient use of the dc voltage in distribution of electrical energy.

1.1.4 DC Versus AC

In both power transmission and distribution, the use of dc voltages and currents offers a number of technical and economical benefits. Therefore, the question arises whether the ac grids would gradually disappear, and if so, when is it going to happen? It has to be noted that the electrical power system is one of the largest systems made by man. The relevant equipment and systems have considerable value, and their installation requires significant investment. Once the equipment is installed and running, there are good reasons to keep it in use throughout the expected service life. At the same time, interruption of dc currents is more difficult, while contemporary power electronics devices have a rather limited peak currents. For this reason, the expansion of dc transmission grids and dc distribution grids is gradual, and it could take considerable amount of time for the ac grid to disappear altogether. Thus, it is reasonable to expect simultaneous use of both ac and dc grid for prolonged periods of time.

1.2 Topologies and Functionality

Putting aside the single-phase converters and dc/dc converters in microgrids, most grid-side power converters are the three-phase inverters that provide the power interface between the dc-bus and the ac grid with symmetrical three-phase voltages. Generic 0.4 kV inverters employ the switching bridge with six semiconductor power switches. In most cases, control of the power switches is based on the pulse width modulation, wherein the voltage pulses are changed in order to create the sinusoidal change of the average output voltages in each phase, thus matching the voltages of the ac grid. Due to the pulsed nature of the inverter output, it is necessary to introduce an *LCL* filter (Fig. 1.1) that suppresses the PWM ripple and provides smooth, sinusoidal line currents.

1.2.1 Medium- and High-Voltage Converters

Due to limited breakdown voltage of semiconductor power switches, the grid-side power converters for medium and high voltages require advanced topologies, such

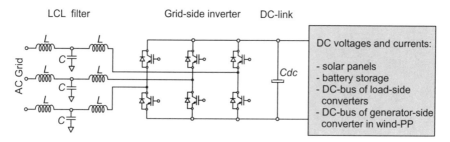

Fig. 1.1 Grid-side inverter in 0.4 kV ac grids

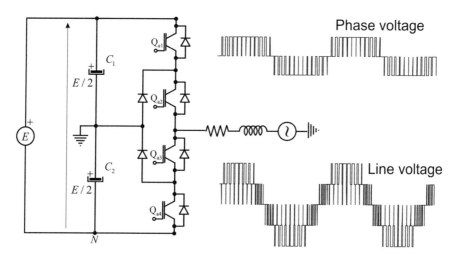

Fig. 1.2 One phase of three-level converter

as multilevel inverters and modular multilevel converters. One phase of the three-level inverter is shown in Fig. 1.2. The voltage across each of IGBT switches is one half of the bus voltage E, which allows the operation with dc-bus voltages larger than the breakdown voltage of semiconductor devices. The phase voltage has three voltage levels ($-E/2$, 0 and $+E/2$), while the line-to-line voltage has five levels ($-E$, $-E/2$, 0, $+E/2$ and E), thus reducing the corresponding ripple.

Converter stations that connect high-voltage dc (HVDC) transmission lines to ac transmission lines are mostly designed as modular multilevel converter (MMC) topologies [1–3]. In each phase, there are several series-connected lower-voltage cells (Fig. 1.3) which can generate considerably larger output voltages (V_1, V_2, V_3 in Fig. 1.3). Practical implementations of MMC topology in HVDC converters have six chains of series-connected H-bridge cells, organized in three phases. In each phase, one chain connects the phase output to the plus rail of the dc-bus, while the other connects the phase output to the minus rail of the dc-bus. Namely, the series-connected H-bridge chains replace the IGBT transistors of Fig. 1.1.

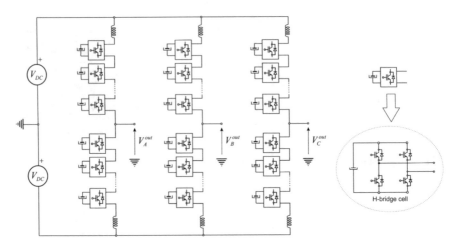

Fig. 1.3 Basic principle of multilevel multicell converters

1.2.2   The Voltage Control in Grid-Side Converters

The source-side converters are often used instead of traditional synchronous generators, which provide the three-phase symmetrical system of electromotive forces connected to the grid across the equivalent series inductances. Similarly, the source-side power converters could be conceived as the three-phase symmetrical ac voltage sources connected to the grid across the *LCL* filter. In cases where it is necessary to provide a virtual synchronous machine, namely, to emulate the behavior of an actual synchronous machine, the *LCL* filter acts as the series inductance, while the PWM-generated voltages have to act as an electromotive force. The voltage actuators in contemporary grid-side converters are mostly PWM-controlled voltage source inverters with semiconductor power switches (Fig. 1.1). The output voltages should be sinusoidal, with their frequency equal to the line frequency and with the amplitude and phase determined so as to create the desired active and reactive power. The switching structures of Figs. 1.1, 1.2, and 1.3 generate pulsating voltages, and they cannot provide the sinusoidal output. With sinusoidal change of the pulse width, the average voltage in each phase has a sinusoidal change. The switching frequency component of the pulsed voltages is removed by the *LCL* filters.

1.2.3   Current Control in Grid-Side Converters

The purpose of digital current controllers is to control the electrical current injected into the grid. The goal of the current control is the suppression of any error between the current reference and the actual, measured current, obtained from the current sensors. An ideal current controller has to provide an error-free tracking of the current reference profiles even in the presence of the input disturbances (the changes of the reference) and the voltage disturbances (the changes of the grid voltages). The driving force of the current controller are the voltages generated by the switching bridge through the PWM process.

The current controller considers the current error and calculates the voltage required to drive detected error back to zero. The controller outputs such voltage as a signal called the voltage reference. Practical implementation of current controllers is digital, and it involves the PWM voltage actuator, the current sensors, the feedback acquisition system that turns analogue signals into digital, and the digital, discrete-time controller. The current controllers are used in grid-side inverters but also in ac drives. In both applications, the model of the load is similar, and it comprises the series impedance and a back-electromotive force (Fig. 1.2). The line voltage of ac grid corresponds to the back-electromotive force of ac machines, while the resistance and the equivalent series inductance of the ac machine windings correspond to the series resistance and inductance placed between the grid-side inverter terminals and the connection to the ac grid.

In both the grid-side inverters and ac drives, the steady-state output currents are ac currents. Their frequency is either the line frequency of the ac grid or the frequency that corresponds to the speed of the revolving magnetic field within ac machines. Due to finite gains and a finite closed-loop bandwidth of the current controller, the operation with ac current references introduces the phase and amplitude errors. In order to obtain an error-free current control of line-frequency ac currents, it is necessary to transform the voltage and current vectors in d-q coordinate frame, where the steady-state values of the currents are dc quantities, and the presence of an integral gain ensures the steady-state error equal to zero.

1.3   The Impact of Grid-Side Converters

Due to imperfections of semiconductor switches, sensors, and controls, the grid-side power converters could inject low-order harmonics and small dc bias components into the ac grid and also introduce stability problems.

1.3.1   DC Bias and Line Harmonics

The principal components of grid-side inverters are electronically controlled switching bridges with semiconductor power switches. The gate drivers and control circuitry have to employ a certain lockout time. The lockout time-related voltage errors contribute to the difference between the output voltages and the sinusoidal pulse width commands, that is, the modulation signals. The voltage error caused by the lockout time comprises significant amount of low-order harmonics. In conjunction with relatively low series inductances, the lockout time can cause a considerable amount of low harmonics of the output current.

The lockout time error can be predicted and compensated. Yet, this compensation cannot be perfect, due to uncertainty of time delays of the gating signals and due to temperature dependence of semiconductor devices. At the same time, the closed-loop current controller reduces the low-order current harmonics, but again, their attenuation cannot be perfect. Eventually, a certain amount of low-order harmonics gets injected into the grid. They have harmful effect on the loads, sources, and devices connected to the grid. Therefore, it is of interest to devise the control means of reducing the content of low-order harmonics. Moreover, there is benefit in devising the control means that would turn each grid-side converter into the agent that absorbs the low-order line harmonics, whatever their origin might be.

The grid-side inverters can also inject undesirable, parasitic dc component into the grid. The voltages are generated by the pulse width modulation, and the current control relies on dedicated current sensors. There is no intention to inject dc current into the ac grids. Yet, due to imperfection of sensors, actuators, and controllers, it is possible for the grid-side inverter to introduce not only the ac currents into the grid but also a small, parasitic amount of dc current, usually called the dc bias. Injection of dc currents jeopardizes the operation of line-frequency power transformers and other devices. In addition to that, the presence of dc bias can produce adverse and harmful consequences in many electrical loads. With an increased number of grid-side power converters, negative consequences of the dc bias are emphasized ever-more. For this reason, the dc bias has to be measured with a very high precision, providing at the same time the means for active compensation and absorption of the bias. Namely, with the proper measurement of the bias, the nearby grid-side converters can provide the counter-injection that would compensate the bias and bring the parasitic dc voltage to zero.

1.3.2 Behavioral Model of Load-Side Converters

The currents and voltages in grid-side power converters are electronically controlled, with the current, power, and dc-bus voltage loops implemented in a digital, discrete-time manner. Considering an electrical load such as the electrical drive with an active front-end converter that interfaces the grid, the motor-side inverter resides in the shaded box in the right side of the dc-bus in Fig. 1.1. The grid-side converter (in the left side of Fig. 1.1) controls the line currents and converts the power of ac voltages and currents into the power injected into the dc-bus, suited to balance the load power, thus keeping the dc-bus voltage in balance. Neglecting the losses, the grid ac power is determined by the load power. In the case of an electrical drive, it is defined by the motor speed and torque.

The load power of electronically controlled loads does not depend on the line voltages. With constant load power, the grid ac power remains constant in the presence of changes in ac voltages. Whenever the line voltages increase, the grid-side converter currents would decrease in amplitude in order to maintain the desired power. The change of the current amplitude comes from the internal controller which maintains the dc-bus voltage. Namely, with an increase of the grid voltages, and assuming that the line-current amplitude remains initially unchanged, the power of the grid-side converter increases and becomes larger than the load power. In consequence, the dc-bus capacitor gets charged, and the dc-bus voltage increases. Further on, the internal controller which maintains the dc-bus voltage reduces the amplitude of the line-current references, and the grid-side power settles back to the steady-state value. In a like manner, whenever the line voltages decrease, the line-current amplitude increases.

From the above considerations, it is concluded that the electronic control of the converter structure of Fig. 1.1 contributes to the grid-side converter behavior which includes negative equivalent resistance. Namely, any positive change ΔU of the grid voltages causes a negative change ΔI of the line currents and vice versa. In other words, the equivalent dynamic resistance of the grid-side converter is negative, $R_{EQ} = \Delta U / \Delta I < 0$. Negative resistance within the behavioral model reduces the stability margin of the ac grid. In cases where the grid has a large number of electronically controlled loads, their negative dynamic resistance can bring the system to instability. In order to avoid the problem, it is necessary to add the means for the local energy accumulation and to use to aid keeping the line current proportional to the voltage transients over the initial, short intervals of time, thus avoiding the negative resistance effects in the frequency range that are critical for the stability of the grid.

1.3.3   Behavioral Model of Source-Side Converters

The source-side converters replace the conventional synchronous generators. Static and dynamic characteristics of synchronous generators were the backbone for the development of many control and protection mechanisms of contemporary ac grids. In this initial phase of deployment of electronically controlled sources, it is desirable that they mimic the static and dynamic characteristics of synchronous machines.

Conventional synchronous generators comprise the three-phase stator winding, the excitation winding on the rotor, and the damping winding on the rotor. Subtransient dynamic processes of the damping winding decay in tens of milliseconds. They are followed by transient dynamic processes within the excitation winding, which decay in several hundred of milliseconds. Both processes have considerable impact on the generator capability to supply the fault currents in the case of the short circuit, and they have the key role in designing and tuning the ac grid protection mechanisms. The capability of grid-side converters to emulate the short-circuit behavior of synchronous generators is limited by the current carrying capability of semiconductor power switches.

The power delivered by the synchronous generator changes with the line voltage, with the rotor electromotive force, and with the winding inductances, and it depends on $\sin(\delta)$, where δ is the angle between the rotor electromotive force and the stator voltage. In steady state, the rotor electromotive force and the stator voltage revolve at the same speed, and the angle δ is constant. In transient conditions, the rotor acceleration can introduce small, transient differences between the rotor speed and the line angular frequency. When the rotor speed increases, the angle δ increases as well, as well as the power delivered by the generator. In cases where the rotor speed transients fall below the line angular frequency, the angle δ and the generated power both decrease.

The change of the generator power, the rotor speed, and the angle δ are described by a second-order differential equation. The damping factor of the oscillatory response is determined by the damper winding, which introduces an additional component of the electromagnetic torque. This new component exists only in transient states. It is proportional to the slip, namely, to the transient difference between the rotor speed and the line angular frequency. Any change of the phase and frequency in the ac grid provokes the transient changes of generated power in all the synchronous generators connected to the grid. It is desirable that the source-side converters mimic this change. In cases such as the solar power plants and the wind power plants, the power delivered to the ac grid depends on the energy harvesting, that is, it depends on weather conditions, the solar power, and the wind speed. Any request to introduce intermittent, short-term transient changes of the electrical power requires some means of local energy accumulation that can provide the transient power peaks that could last several seconds.

1.4   Control Techniques for Grid-Side Converters

Control tasks of grid-side power converters include digital current control; suppression and absorption of parasitic dc bias; suppression of line harmonics; control of active and reactive power; supply of controllable grid-fault currents; frequency, voltage, and droop control; emulation of synchronous generators; and other tasks.

1.4.1   Robust and Error-Free Feedback Acquisition

The pulse width modulated voltage pulses obtained at the output of the three-phase inverter comprise the frequency components at the switching frequency $fPWM$ and above. The output current comprises the frequency component called the current ripple, which has the components at the same frequencies as the output voltage. At the same time, the sampling frequency of digital current controller is determined by the PWM frequency. For this reason, sampling of the feedback signals (the output currents) introduces the sampling errors. The errors take place in cases where the analogue input signal comprises the frequency components above one half of the sampling frequency. The sampling and reconstruction of analogue signals are discussed by Kotelnikov and Shannon [4–6]. For this reason, the feedback acquisition system has to remove the signals at the PWM frequency and its multiples.

In most cases, the undesired frequency components at higher frequencies are removed by means of analogue low-pass filters, also called the anti-aliasing filters. The theory requires complete removal of any frequency component above one half of the sampling frequency. In practical applications with limited resolution of A/D converters, it is sufficient to suppress the undesired frequency components below the level of one least significant bit of the A/D converter (1 LSB, the smallest detectable change of the analogue signal). Even so, the anti-aliasing low-pass filter with desired attenuation introduces considerable phase and amplitude errors at frequencies lower than one half of the sampling frequency and even at frequencies below the desired bandwidth. Therefore, it is necessary to devise the means of removing the harmful frequency components from the train of feedback samples while maintaining the phase and amplitude of the feedback signal unaltered from dc up to the bandwidth frequency.

Error-free ripple-insensitive sampling can be achieved by sampling the feedback signal at 2^N times larger sampling frequency, thus acquiring a large number of equidistant samples within each PWM period. The feedback signal is obtained by averaging the train of samples acquired within the past PWM period. In this way, all the components at the PWM frequency and its multiples are removed from the feedback chain. The implementation of this oversampling-based one-PWM period feedback averaging relies on advanced features of digital signal controllers. Fast A/D acquisitions are performed each 2–3 μs by means of an autonomous sequencing machines. At the same time, data collection and storage are performed without CPU load, relying on an internal DMA machine which handles automated data transfers between the peripherals and RAM.

1.4.2   High Bandwidth Digital Current Controllers

The innermost loop in grid-side power converters is the digital current loop. Performances of the inner loop greatly affect the overall behavior of the converter. The principal component of grid-side inverters are electronically controlled switching bridges with semiconductor power switches. The task of the digital current controller includes the acquisition of the feedback signals, discrimination of the current error, calculation of the voltages that are suited to suppress the current error, and implementation of these voltage commands through the PWM process.

The digital current controllers are of discrete-time nature. The analysis and design of discrete-time current controllers rely on z transform, and it takes into account all the transport delays caused by the feedback acquisition, computation, and PWM processes. Design and parameter setting of the discrete-time controller has to include all the delays in the feedback path and all the delays in the PWM path. One PWM period averaging improves the robustness of the controller against the noise in the feedback path, but it also introduces an additional transport delay in the closed-loop chain. In order to obtain a quick, well-damped, and robust response, it is necessary to design the controller using the internal model control (IMC) approach and to set the parameters to maximize the closed-loop bandwidth yet preserving an overshoot-free response and sufficient robustness against the parameter changes. The limiting factor for the closed-loop bandwidth are the transport delays. The effect of delays can be reduced by the proper scheduling of the control tasks within each interrupt and by introducing the series compensator with differential action.

1.4.3   Suppression of Low-Order Harmonics

Digital current controllers have the crucial impact on performance of grid-side converters and ac drives. In addition to an error-free tracking of the input reference, the task of the current controller is also the suppression of the voltage disturbance. In grid-side inverters, the voltage disturbances are the line voltages. In ac drives, the voltage disturbance are the back-electromotive forces of ac machines. The voltage disturbances are commonly suppressed by enhancing the controller with an additional inner loop, the active resistance feedback. It is implemented as an additional proportional gain. Suppression of the voltage disturbances is proportional to the active resistance gain. Therefore, it is of interest to maximize that gain. The time delay introduced into the feedback path creates difficulties in designing the current controller with the active resistance. In order to increase the active resistance gain, it is necessary to analyze the impact of transport delays and to provide the compensation schemes.

1.4.4   Synchronization and Power-Frequency Change

Typical source-side converters are the inverters that collect the electrical energy from the wind power plants or solar power plants, convert the energy into a set of three-phase voltages and currents, and inject the active and reactive power into the three-phase ac grid. With the advent of local accumulation and considering the regeneration needs of electrical drives, most load-side converters have to be bidirectional, capable of supplying electrical energy into the grid during brief intervals of time. Therefore, the basic functionality of all the grid-side converters is similar. When interfacing the ac grids, the grid-side power converter has to provide the voltages and inject the currents that are in synchronism with the grid voltages. Therefore, it is necessary to provide the means for detecting the frequency and the phase of the grid ac voltages.

Most common device in use is the phase-locked loop (PLL), often used in radio circuits. Dynamic properties of the grid synchronization device have a significant impact on the response of the grid-side converter power to the grid transients. With the present infrastructure, the power response of the grid-side converter to the grid transients should, ideally, resemble the response of a typical synchronous generator. An increased share of grid-side power converters with different dynamic properties could have adverse effects on the grid operation and stability. Therefore, it is of interest to study the grid synchronization devices and to find the way to make their dynamic properties closer to those of synchronous generators.

References

1. Gowaid IA, Adam GP, Massoud AM, Ahmed S, Holliday D, Williams BW (2014) Quasi two-level operation of modular multilevel converter for use in a high-power DC transformer with DC fault isolation capability. IEEE Trans Power Electron 30(1):108–123
2. Montesinos-Miracle D, Massot-Campos M, Bergas-Jane J, Galceran-Arellano S, Rufer A (2013) Design and control of a modular multilevel DC/DC converter for regenerative applications. IEEE Trans Power Electron 28(8):3970–3979
3. Deng F, Chen Z (2014) A control method for voltage balancing in modular multilevel converters. IEEE Trans Power Electron 29(1):66–76
4. Kotelnikov VA (1933) On the capacity of the "ether" and of cables in electrical communication. In: Proceedings of the first all-union conference on the technological reconstruction of the communications sector and low-current engineering, Moscow
5. Shannon CE (1948) A mathematical theory of communication. Bell Syst Tech J 27:379–423. 623–656
6. Kotelnikov VA (1959) The theory of optimum noise immunity. McGraw-Hill, New York/Toronto/London

Chapter 2
PWM Voltage Actuator

Digital current controllers are used in grid-side inverters, generator-side inverters, electrical drives, and many other applications. Their purpose is to control the electrical current injected into the grid or the electrical current supplied to the windings of an electrical machine. Desired goals include quick and accurate suppression of any errors between the desired current, also called the current reference, and the actual current obtained from the current sensors. Ideally, digital current controllers should provide the means for an error-free tracking of the current reference profiles even in the presence of the input disturbances (i.e., sudden changes of the reference) and the voltage disturbances (sudden changes of the line voltages, back-electromotive forces, and similar). The driving force in current controllers is the voltage supplied to the load. In grid-side inverters, the load is the ac-grid, which is connected across the LCL filter. In electrical drives, the load can be represented by the series inductance and the series resistance of the stator winding, with the stator back-electromotive force closing the circuit. In essence, the current controller uses the current error to calculate the voltages that are required to drive detected error back to zero. The controller outputs the voltage references. It takes the voltage actuator to generate the actual voltage that corresponds to the voltage reference. In other words, the voltage actuator is an amplifier that turns the voltage reference signals into power circuit voltages.

The voltage actuators that are used in contemporary grid-side converters are mostly PWM-controlled voltage source inverters with semiconductor power switches. Two such inverters are used within a wind power setup, as shown in Fig. 2.1.

Ideally, the voltages obtained at the output of the grid-side converter (Fig. 2.1) should be sinusoidal, with their frequency equal to the line frequency and with the amplitude and phase that create the desired active and reactive power. On the generator side, the frequency of sinusoidal voltages is determined by the speed of the wind turbine. Sinusoidal voltages with adjustable frequency, amplitude, and phase can be obtained from class AB linear amplifiers. Yet, the efficiency of such amplifiers is rather low, as their operation produces a great deal of losses. Instead, the

S. N. Vukosavic, *Grid-Side Converters Control and Design*, Power Electronics and Power Systems, https://doi.org/10.1007/978-3-319-73278-7_2

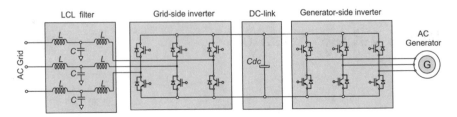

Fig. 2.1 Three-phase IGBT inverters as the voltage actuators in grid-side and generator-side converters of the wind power setup

relevant voltages in Fig. 2.1 are obtained from the switching bridges that provide the train of voltage pulses. The average value of the output voltage is altered (*modulated*) by changing the width of the voltage pulses. The *pulse width modulation* (PWM) technique is introduced in the following section.

2.1   Two-Level Inverters with Symmetrical PWM

A three-phase PWM inverter shown in Fig. 2.2 provides the power interface between the dc source E and the three-phase source of ac voltages (E_a, E_b, E_c). The ac electromotive forces with series R-L impedance can represent the grid as well as a three-phase electrical machine. The plus and minus rails of the dc supply are also called the dc-bus. Any parallel capacitor connected between the rails is called the dc-bus capacitor. For convenience, the dc voltage source E is split in two, and the connecting point between the two $E/2$ sources is taken as the reference (0 V) for the voltage measurements. The voltage at the star connection of the three-phase system is denoted by u_{ST}.

2.1.1   Pulse Width Modulation

There are two semiconductor power switches in each of the three inverter phases. By closing the phase A upper switch (A_H), the output voltage u_a gets equal to $+E/2$. With the lower switch closed (A_L), the output voltage u_a becomes $-E/2$. An attempt to close both A_H and A_L would result in a destructive short circuit. Thus, the phase A output voltage can assume one of the two discrete values, either $+E/2$ or $-E/2$. The instantaneous value of $u_a(t)$ cannot assume a sinusoidal change, but it is possible to use the switching action of A_H and A_L to generate the voltage u_a as a series of pulses (Fig. 2.3).

Assuming that the voltage pulses in Fig. 2.3 repeat with the period T_{PWM} and that the upper switch (A_H) remains closed during the interval t_{ON}, the average voltage u_{av} in one T_{PWM} period changes with t_{ON} in a linear manner:

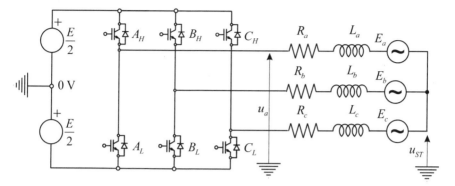

Fig. 2.2 Three-phase IGBT inverter with the three-phase load, approximated by R-L series impedance and an electromotive force

Fig. 2.3 Pulse-shaped phase voltage u_a at the output of the inverter

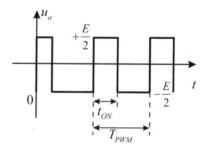

$$u_{av} = \int_0^{T_{PWM}} u_a(t) \cdot dt = \frac{2t_{ON} - T_{PWM}}{T_{PWM}} \cdot \frac{E}{2} = (2m - 1) \cdot \frac{E}{2}. \qquad (2.1)$$

The ratio $m = t_{ON}/T_{PWM}$ changes from 0 ($t_{ON} = 0$) to 1 ($t_{ON} = T_{PWM}$). The average voltage u_{av} in each PWM period changes from $-E/2$ to $+E/2$ while m sweeps from 0 to 1. The pulse width $t_{ON} = mT_{PWM}$ does not have to be the same in the successive PWM periods. Adopting a sinusoidal change of t_{ON}, the voltage u_a becomes a train of voltage pulses with an average value u_{av} that exhibits a sinusoidal change. In addition to the average value u_{av}, the voltage u_a also contains a zero-average ac component that includes pulsations at the frequency $f_{PWM} = 1/T_{PWM}$. Thus, the pulse width modulation (Fig. 2.3) can provide the sinusoidal change of the average u_{av}, but the instantaneous value u_a retains its pulsating nature.

The inductances of the *LCL* filter (Fig. 2.1) and the inductances of the generator winding act as low-pass filter that attenuate the effects of the high-frequency pulses on the current. With an ac excitation $U(j\omega)$ at the angular frequency ω, the current response is $I(j\omega) = U(j\omega)/(j\omega L)$, where L is the equivalent series inductance of the *LCL* filter or the equivalent series inductance of the electrical machine connected to the generator-side inverter of Fig. 2.1. The current $I(j\omega)$ is inversely proportional to the angular frequency. Assuming that $\omega = 2\pi f_{PMW}$ is sufficiently high to make the current $I(j\omega)$ negligible, the pulsed ac component of u_a can be disregarded, resulting in $u_a \approx u_{av}$. In such case, the PWM-controlled inverter (Fig. 2.2) can be considered as the source of controllable sinusoidal voltages.

2.1.2 Pulsed Voltages and the Current Ripple

It is of interest to obtain an insight into the PWM frequency component of the output current which is caused by the pulsating nature of the phase voltages. With floating star connection of the three-phase system (u_{ST} in Fig. 2.2), the ripple calculation gets rather involved. To obtain a quick estimate, the analysis is simplified by assuming that $u_{ST} = 0$. In such case, the voltage pulses of u_a do not affect the currents in other phases, while the voltages of the remaining phases do not affect i_a. Thus, it is possible to consider the single-phase representation given in Fig. 2.4.

In steady-state conditions and with $R_a = 0$, the average voltage across the inductance L_a has to be zero. For that reason, the back-electromotive force E_a has to be equal to $u_{av} = (2\,m-1)E/2$. Within the interval $t_{ON} = mT_{PWM}$ (Fig. 2.4), the voltage across the inductance L_a is $u_{L1} = E/2 - u_{av} = (1-m)E$. Within the interval $t_{OFF} = (1-m)T_{PWM}$, the voltage across the inductance L_a is equal to $u_{L2} = u_{av} + E/2 = mE$. The average voltage across L_a is $t_{ON}\,u_{L1} + t_{OFF}\,u_{L2}$ is equal to zero. Therefore, the current i_a maintains its average value I_{av}. The ac component of the voltage across L_a contributes to linear change of i_a around the average value. The amplitude of i_a oscillations is ΔI. During the interval t_{ON}, the voltage u_{L1} introduces the changes in i_a by the amount of $2\Delta I$, and this relation is used in Eq. (2.2) to

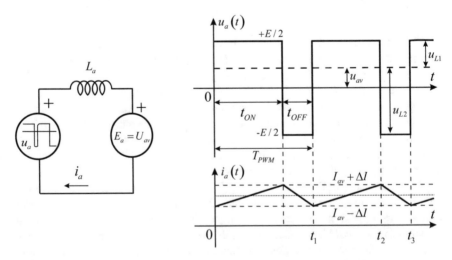

Fig. 2.4 The ripple ΔI of the phase current obtained from Fig. 2.2 with $u_{ST} = 0$

calculate the ripple ΔI in terms of m, E, L_a, and T_{PWM}. The ripple reaches the maximum value ΔI_{max} for $m = 0.5$.

$$u_{L1}t_{ON} = 2L_a\Delta I \Rightarrow \Delta I = \frac{u_{L1}t_{ON}}{2L_a} = \frac{(m - m^2) \cdot E \cdot T_{PWM}}{2L_a},$$

$$\Delta I_{max} = \Delta I\bigg|_{m=0.5} = \frac{E \cdot T_{PWM}}{8L_a}. \tag{2.2}$$

Question (2.1) The three-phase inverter with the rated power of $S_n = 10$ kVA, $E = 600$ V and with $f_{PWM} = 10$ kHz gets connected to the mains voltage of 3×400 V, 50 Hz by means of three series inductances L. Relative (per unit) value of each series inductance is 0.1 (10%). What is the maximum value of the current ripple?

Answer (2.1) The maximum value of the current ripple is calculated from (2.2). The values of E and $T_{PWM} = 1/f_{PWM}$ are available, while it is necessary to calculate L. With $L = 0.1$ p.u, the rated current I_{nrms} produces the voltage drop $\Delta U = 2\pi\, 50\, L\, I_{nrms}$ across the inductance which is equal to 10% of the rated phase voltage, that is, $\Delta U = (0.1)\, 400/1.732 = 23.1$ V. With $I_{nrms} = S_n/400/1.732 = 14.43$ A, $L = 5.1$ mH. From (2.2), $\Delta I_{max} = 1.47$ A, roughly 10% of the rated current.

2.1.3 Star Connection and Line Voltages

The phase voltage of Fig. 2.3 has two discrete levels, $+E/2$ and $-E/2$. The three-phase circuits of Fig. 2.2 have a floating star connection with the voltage u_{ST} which can assume non-zero values. The voltage u_{ST} (Fig. 2.2) can be expressed in terms of the parameters and variables of phase A:

$$u_{ST} = u_a - R_a i_a - L_a\frac{di_a}{dt} - E_a \tag{2.3}$$

In a like manner, the voltage u_{ST} can be expressed in terms of the parameters and variables in the remaining two phases. The three expressions can be summed, resulting in

$$3 \cdot u_{ST} = u_a + u_b + u_c - R_a i_a - R_b i_b - R_c i_c -$$
$$-L_a\frac{di_a}{dt} - L_b\frac{di_b}{dt} - L_c\frac{di_c}{dt} - E_a - E_b - E_c \tag{2.4}$$

The sum of the three-phase currents in Fig. 2.2. is equal to zero, $i_a + i_b + i_c = 0$. In cases where the resistances and the inductances of the three-phase circuit are equal ($R_a = R_b = R_c$, $L_a = L_b = L_c$) and where the back-electromotive forces of Fig. 2.2 make up a symmetrical system with $E_a + E_b + E_c = 0$, the voltage u_{ST} is equal to the average value of the three-phase voltages:

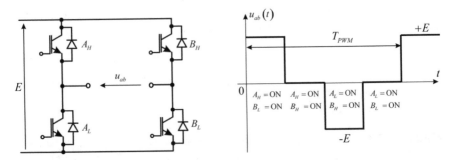

Fig. 2.5 Line-to-line voltage at the output of the inverter: four switching states and three levels

$$u_{ST} = \frac{u_a + u_b + u_c}{3} \qquad (2.5)$$

Thus, the voltage u_{ST} of Fig. 2.2 can float from $-E/2$ ($A_L = B_L = C_L = $ ON) to $+E/2$ ($A_H = B_H = C_H =$ ON). With the star connection floating, the phase currents i_a, i_b, and i_c depend on the line-to-line voltages u_{ab}, u_{bc}, and u_{ca}. While the phase voltages have two discrete levels, the line voltages can be $-E$, 0, or $+E$. The line voltage u_{ab} is given in Fig. 2.5, along with the corresponding switching states. Considering the need to create a sinusoidal change of the line voltage average $u_{ab(av)}$, a convenient solution is using the switching states with $u_{ab} = +E$ and $u_{ab} = 0$, while $u_{ab(av)} > 0$, while reverting to the states with $u_{ab} = -E$ and $u_{ab} = 0$ when $u_{ab(av)} < 0$. However, the switching states in phases A and B also affect the line voltages u_{bc} and u_{ca}, and this has to be taken into account when devising the switching strategy suitable for the three-phase pulse width modulation.

Question (2.2) Considering the phase voltage u_a, generated according to the PWM pattern illustrated in Fig. 2.3; observing at the same time the relation between the pulse width t_{ON} and the average value $u_{a(av)}$ of $u_a(t)$ within each T_{PWM} period; neglecting the pulsating components of $u_a(t)$ at the switching frequency, and assuming that the pulse width t_{ON} changes so as to obtain a slow, sinusoidal change of the average value $u_{a(av)}$ at the line frequency, calculate the largest rms value of the sinusoidal change of $u_{a(av)}$.

Answer (2.2) The average value of $u_{a(av)}$ changes between $-E/2$ ($\underline{A}_L = $ ON) and $+E/2$ ($A_H =$ ON). Thus, any sinusoidal change of $u_{a(av)}$ remains within the interval $[-E/2 .. +E/2]$. Therefore, the maximum peak value of the sinusoidal change is $u_{amax} = E/2$, while the rms value of the sinusoidal $u_{a(av)}$ change is $U_{arms1} = E/2/\text{sqrt}$ (2).

Question (2.3) Assuming that the phase voltages u_a, u_b, and u_c are generated as described in the previous question and that the phase shift between phase voltages is $2\pi/3$, what is the rms value of the sinusoidal change of line voltages?

Answer (2.3) In a symmetrical three-phase system, the line voltages are 2 sin($2\pi/$ 3) $=$ sqrt(3) times larger than the phase voltages. Therefore, the required rms value is $U_{ab\ rms1} = E$ sqrt(3/2)/2.

Question (2.4) Considering the line voltage u_{ab}, generated according to the PWM illustrated in Fig. 2.5, and assuming that the ac components of $u_{ab}(t)$ at the switching frequency are of no significance, calculate the rms value of the sinusoidal change of $u_{ab(av)}$, where the latter is the average value of $u_{ab}(t)$ within each T_{PWM} wide switching period.

Answer (2.4) The average value of $u_{ab(av)}$ in Fig. 2.5 sweeps between $-E$ ($A_L = B_H = $ ON throughout the switching period) and $+E$ ($A_H = B_L = $ ON). Thus, any change of $u_{ab(av)}$ remains within the interval $[-E .. +E]$. The maximum peak value of the sinusoidal change is $u_{ab\ max} = E$, while the corresponding rms is $U_{ab\ rms2} = E/\text{sqrt}(2)$. Notice at this point that $U_{ab\ rms2}$ is 2/sqrt(3) times larger than $U_{ab\ rms1}$ obtained in Question 2.3.

There are six power switches in a three-phase PWM inverter (Fig. 2.2). Their state is controlled by the control electrode. With MOSFET and IGBT transistors, the voltage at the control electrode (the gate) controls the state of the switch. Whenever the voltage between the gate and the emitter (V_{GE}) exceeds the threshold of 5–6 V, the IGBT transistor is in conduction. Therefore, it is a common practice to maintain $V_{GE} = -15$ V in order to keep the switch OFF and to supply $V_{GE} = -15$ V in order to turn the switch ON. The gate voltages are supplied by dedicated electronic circuits called *the gate drivers*. Two such drivers are shown in Fig. 2.6. Pulse width modulated signal is obtained from the peripheral unit of the microcontroller. The unit comprises programmable timers that generate the pulses with period T_{PWM} and the pulse width t_{ON}. While the output of the peripheral unit refers to the microprocessor ground, the drivers are tied to the power circuits and, therefore, exposed to dangerous potentials. For this reason, digital PWM signal cannot be directly wired to the drivers. Instead, it has to be passed through the stage with galvanic isolation. The isolation can be accomplished by using the photo-couplers or fiber-optic cables.

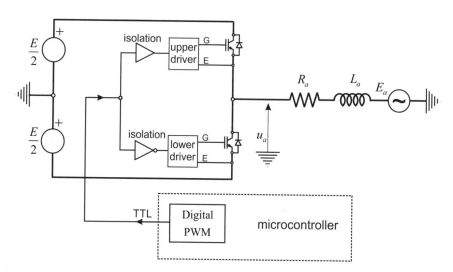

Fig. 2.6 The state of the power switches is controlled from a peripheral unit of the microcontroller which comprises programmable timers

2.1.4   Symmetrical and Asymmetrical PWM Carrier

Desired voltage u_a is often called the voltage *reference* and denoted by u_a^*. In order to obtain $u_{a(av)} = u_a^*$, the pulse width t_{ON} (2.1) should be set to the value calculated in (2.6). The value of u_a^* changes the pulse width in a linear manner, and it is also called the *modulation signal*. Within each PWM period, there is only one voltage pulse and only one, discrete value of t_{ON}. Therefore, it is inappropriate to represent the modulation signal with a continuous-time function $u_a^*(t)$. Instead, the modulation signal can be represented as a train samples, implying a single, discrete value u_a^* (nT_{PWM}) that corresponds to the nth switching period. In cases where the desired phase voltage has a sinusoidal change with the fundamental frequency ω_F, the value $u_a^*(nT_{PWM})$ within the interval $[nT_{PWM} .. (n + 1)T_{PWM}]$ should be equal to $U_m \sin(\omega_F\, n\, T_{PWM})$. Corresponding pulse width is given in (2.7).

$$t_{ON} = \frac{T_{PWM}}{2}\left(1 + u_a^* \frac{2}{E}\right). \tag{2.6}$$

$$t_{ON}^n = \frac{T_{PWM}}{2}\left(1 + \frac{2U_m}{E}\sin(\omega_F \cdot n \cdot T_{PWM})\right). \tag{2.7}$$

$$m^n = \frac{t_{ON}^n}{T_{PWM}} = \frac{1}{2} + \frac{U_m}{E}\sin(\omega_F \cdot n \cdot T_{PWM}). \tag{2.8}$$

The ratio $m = t_{ON}/T_{PWM}$ is often called the modulation index, and it is given in (2.8). In cases where the switching frequency exceeds the fundamental frequency by several orders of magnitude, the pulsed nature of the voltage $u_a(t)$ can be neglected along with the current ripple (2.2). With $\omega_F \ll 2\pi T_{PWM}$, the modulating signal can be approximated by the function $u_a^*(t)$ which coincides with the actual train of samples, while the inverter can be envisaged as an amplifier that provides the output voltage $u_{a(av)}(t) = u_a^*(t)$.

The process of pulse width modulation in Fig. 2.7a involves the modulation index m and the PWM carrier signal $c(t)$. The signal $c(t)$ in Fig. 2.7a has a sawtooth form. The output voltage $u_a(t)$ remains $+E/2$ until the intersection of the modulation signal m and the sawtoothed carrier $c(t)$, when it commutates and remains $-E/2$ until the end of the PWM period. The voltage pulses are not symmetrical with respect to the PWM periods. Therefore, the process in Fig. 2.7a is known as *asymmetrical* PWM. When the modulation index m sweeps from 0 to 1, the pulse width t_{ON} changes from 0 to T_{PWM}, and the average value of the output voltage changes from $-E/2$ to $+E/2$.

The PWM carrier in Fig. 2.7b has a triangular shape. The impact of the modulation index m on t_{ON} and on $u_{a(av)}$ remains the same as with asymmetrical PWM. Due to triangular shape of $c(t)$, the center of the negative voltage pulse coincides with the peak of the PWM carrier. In cases where $m_{n-1} = m_n$, the center of the positive voltage pulse coincides with the zero point of the PWM carrier.

Symmetrical placement of the voltage pulses with respect to the triangular carrier is the reason to call the modulation scheme of Fig. 2.7b *symmetrical* PWM.

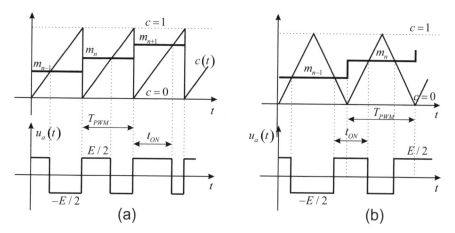

Fig. 2.7 Line-to-line voltage at the output of the inverter: four switching states and three levels

Symmetrical PWM is used in most grid-side power converters and electrical drives due to its advantages over the asymmetrical PWM.

An important drawback of asymmetrical PWM can be observed by considering the three-phase PWM inverter and the instant when the PWM carrier $c(t)$ in Fig. 2.7a drops from one to zero. At that instant, the phase voltages u_a, u_b, and u_c commutate from $-E/2$ to $+E/2$. The commutation takes place in all the three phases, and it contributes to commutation losses. At the same time, the line-to-line voltages u_{ab}, u_{bc}, and u_{ca} do not change, their value was zero before the commutation, and they remain at zero immediately after the commutation. Thus, the energy investment in three commutations does not contribute to any change in the output voltages u_{ab}, u_{bc}, and u_{ca}. This drawback is eliminated by symmetrical PWM in Fig. 2.7b.

With both asymmetrical and symmetrical PWM schemes, the voltage reference affects the output voltage with delay. Namely, the modulation signals in Fig. 2.7 have to be set at the beginning of the PWM period, and they control the average voltage until the end of the relevant PWM period. On an average, the voltage reference is delayed by $T_{PWM}/2$. This transport delay affects the design and the performances of the digital current controllers.

2.1.5 Double Update Rate

The modulation signal in Fig. 2.7b is changed in each PWM period. The sinusoidal change of the phase voltages can be obtained by changing the modulation signal as suggested in (2.8). Each voltage pulse in Fig. 2.7b has its rising edge coinciding with the falling slope of carrier and its falling edge coinciding with the rising slope of the carrier. The average voltage depends on the position of both commutations. Thus, it is possible to use one modulation index to set the rising edge of the voltage pulse

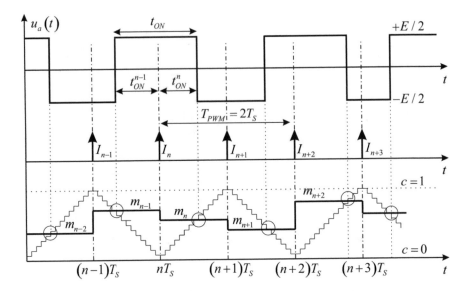

Fig. 2.8 Double-rate symmetrical PWM with corresponding phase voltage $u_a(t)$

while using a different modulation index for the falling edge. Each of these new modulation indices has to be maintained over one half-period T_S of the PWM, $T_S = T_{PWM}/2$. In other words, it is possible to double the update rate of the modulation index, reducing in this way delays in applying the reference voltage at the output terminals of the inverter.

In Fig. 2.8, the basic principles of doubling the update rate are illustrated by showing the symmetrical PWM carrier, the modulation indices set in specific half-periods, and the waveform of the resulting phase voltage. The modulation index m_n has to be set at the instant nT_S, at the beginning of nth half-period T_S. It gets compared to the rising slope of the triangular PWM carrier, and it sets the commutation at instant $nT_S + t_{ON}{}^n$. By affecting $t_{ON}{}^n$, the modulation index m^n determines the average value of the output voltage on the interval $[nT_S .. (n + 1)T_S]$. In a like manner, the modulation signal $m^{n + 1}$ compares to the falling slope of $c(t)$ and controls the average value of the output voltage on the interval $[(n + 1)T_S .. (n + 2) T_S]$. In the described way, delay in applying the reference voltage is halved. On an average, the voltage reference is delayed by $T_S/4 = T_{PWM}/4$. The change of the modulation indices twice per PWM period is usually called *double update rate*, and it is used in most digital current controllers.

2.1.6 The Output Voltage Waveform and Spectrum

The phase voltage at the output of the PWM inverter (Fig. 2.2) comprises the desired sinusoidal change of the average voltage $u_{a(av)}$ in each period. This change represents the desired phase voltage at the fundamental frequency ω_F. In grid-side inverters, ω_F

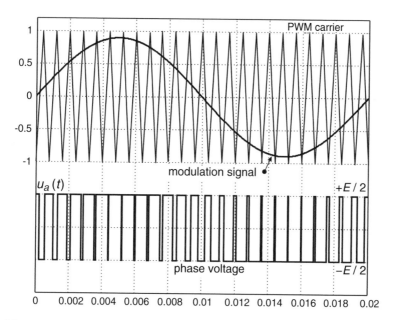

Fig. 2.9 PWM carrier, modulation signal, and the phase voltage $u_a(t)$ with symmetrical pulse width modulation

corresponds to the line frequency, while in electrical drives it depends on the drive speed. Due to pulsed nature of the actual phase voltage, it also comprises the high-frequency components related to the commutation frequency f_{PWM}. It is of interest to get an insight into the actual waveform of the phase voltage and its spectrum.

The waveforms in Fig. 2.9 represent the PWM carrier, the phase voltage $u_a(t)$, and the continuous-time function that represents sinusoidal modulation signal. The waveforms are obtained for the symmetrical PWM with double update rate and with the sinusoidal voltage reference with the voltage amplitude set to 90%. The corresponding spectrum of the phase voltage is given in Fig. 2.10. At $f = f_F = \omega_F/2\pi$, the spectrum has the fundamental component, representing the sinusoidal change of the voltage average value. The spectrum comprises a pronounced component at f_{PWM}. The two visible side components are placed at the distance $\pm 2f_F$. In addition to that, the spectrum has another two groups placed next to $2f_{PWM}$ and next to $3f_{PWM}$. There are also spectral components with minor amplitude, as well as the components in the frequency range above $3f_{PWM}$. All the components above $f = f_F$ are caused by the PWM nature of the phase voltages.

Although the voltage spectrum has considerable high-frequency components, their impact on the phase current is reduced by the low-pass nature of the series impedance. That is, the frequency component of the output voltage $U(j\omega)$ produces the response of the output current described by $I(j\omega) = U(j\omega)/(j\omega L + R)$. Neglecting the resistance, the current response to the voltage excitation is inversely proportional to the excitation frequency.

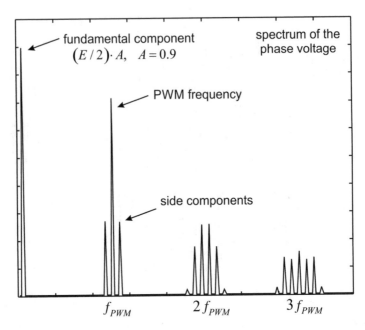

Fig. 2.10 The spectrum of the phase voltage $u_a(t)$ obtained with symmetrical pulse width modulation

The waveforms in Figs. 2.11 and 2.12 correspond to the three-phase PWM inverter which operates with 90% of the output voltage at $f_F = 50$ Hz. The output is loaded with R-L load characterized by $\tau = L/R = 1$ ms. The waveforms are obtained by computer simulations (model1.zip). In Fig. 2.11, it is assumed that the star point is connected to the zero point (Fig. 2.2). Thus, the current in phase A does not get affected by commutations in other phases. The relative value of the switching frequency component within the phase current is considerably lower than the relative value of the switching frequency component within the phase voltage. This illustrates the low-pass nature of series-connected inductive loads and provides the grounds for the use of PWM technique in achieving almost sinusoidal output currents.

Similar test is outlined in Fig. 2.12, where the star point is left floating, with $i_a + i_b + i_c = 0$. The current of the three-phase system with floating star connection depends on the line-to-line voltages. Thus, the current in phase A gets affected not only by the phase voltage u_a but also by commutations in other two phases. This alters the waveform of the current ripple, reducing the apparent peaks and emphasizing the frequency components at higher frequencies.

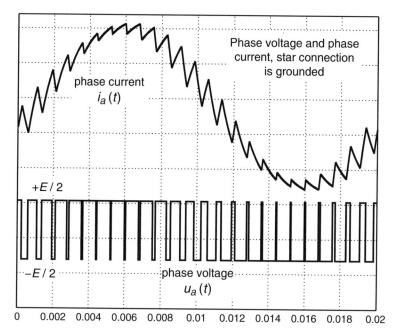

Fig. 2.11 The phase current and the phase voltage obtained at $f_F = 50$ Hz, with $U_m/U_{max} = 0.9$, $L/R = 1$ ms and with the star connection grounded

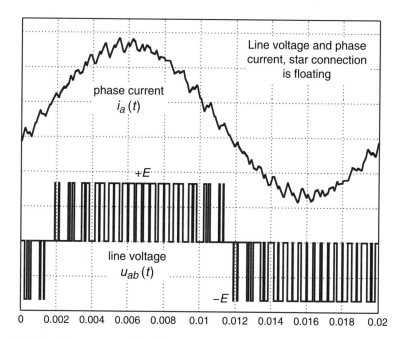

Fig. 2.12 The phase current and the line voltage obtained at $f_F = 50$ Hz, with $U_m/U_{max} = 0.9$, $L/R = 1$ ms and with the star connection floating

2.2   Space Vector Modulation with DI and DD Sequences

The PWM techniques described in Sect. 2.2.1 are focused on controlling a pair of switches in one inverter phase. Instead, it is possible to consider the three-phase inverter as a whole and to devise coordinated control of the switches in all the inverter phases. By doing so, it is possible to increase the available output voltage and to reduce the commutation losses.

The inverter with three phases and six switches has a limited number of switching states. The output voltages of the inverter are u_a, u_b, and u_c, and they can be represented by one voltage vector. Each of the switching states provides one discrete voltage vector, called the space vector. Desired (or reference) voltage vector rarely coincides with one of the discrete voltage vectors. The space vector modulation technique consists in devising a cycle of several discrete voltage vectors and assigning the appropriate dwell times in order to achieve the average voltage equal to the reference.

A three-phase inverter of Fig. 2.2 supplies the phase voltages with their average-per-PWM-period value $u_{a(av)}$ changing from $-E/2$ ($t_{ON} = 0$) and $+E/2$ ($t_{ON} = T_{PWM}$). Presumed sinusoidal change of $u_{a(av)}$ remains within $-E/2$ and $+E/2$ limits. Thus, the largest rms value of the sinusoidal $u_{a(av)}$ change is $U_{arms1} = E/2/sqrt(2)$, and the largest rms value of corresponding line-to-line voltages is $U_{ab\ rms1} = E\ sqrt(3/2)/2$. At the same time, the line voltage waveforms in Figs. 2.5 and 2.12 provide the evidence that the peak value of the line voltage reaches E, thus resulting in $U_{ab\ rms2} = E/sqrt(2)$, the value larger $2/sqrt(3)$ times.

The space vector modulation technique discussed henceforth provides coordinate control in the three inverter phases and allows the line voltage rms to reach $E/sqrt(2)$. The space vector modulation technique provides sinusoidal line voltage, and it is suitable for the three-phase systems with floating star connection or delta connection. Namely, it will be shown that the corresponding phase voltages are not sinusoidal, thus excluding the use in three-phase systems with grounded star connection.

2.2.1   The Switching States and the Voltage Vectors

The three-phase inverter of Fig. 2.2 has six semiconductor power switches, organized in three inverter arms, each arm comprising an upper and a lower switch. At any instant, only one switch is in conduction state. In an attempt to turn on both the upper and the lower switch, the dc-bus voltage would be short-circuited, and the short-circuit current would damage the switches. Moreover, it is not possible to keep both switches off, unless the output is disconnected and the output current is zero. Even in cases where gate voltage is below the conduction threshold at both switches, a non-zero current will circulate in one of the power diodes that are connected in parallel with power transistors. Thus, there are only two switching states in each phase. When either the upper transistor or the upper diode is in conduction, the

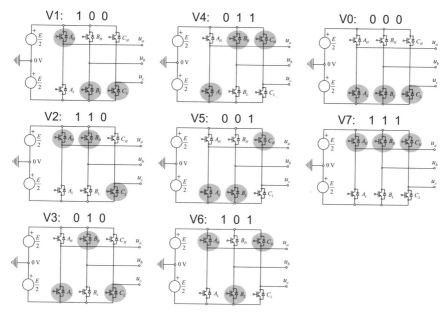

Fig. 2.13 The eight switching states of a three-phase inverter

output voltage is equal to $+E/2$. When either the lower transistor or the lower diode is in conduction, the output phase voltage is equal to $-E/2$. The former can be represented by the switching code "1" and the latter by the code "0." With that in mind, the switching state of the three-phase inverter can be represented by the binary code with three digits, ABC, where any of the digits represents one phase.

The code $ABC = 100$ designates the switching state where the upper switch in phase A is on ($u_a = +E/2$), and the lower switches in phases B and C are also on ($u_b = -E/2$ $u_c = -E/2$). The total number of switching states is $2^3 = 8$. They are illustrated in Fig. 2.13, and each one is given the name ($V_0 .. V_7$).

In systems with floating star connection, any contemporary change in all the phase voltages by the same amount does not affect the output currents i_a, i_b, and i_c, while it changes the voltage u_{ST} of the star connection (2.5). An example is simultaneous change from $u_a = u_b = u_c = -E/2$ to $u_a = u_b = u_c = +E/2$, where u_{ST} changes from $-E/2$ to $+E/2$, while the line voltages u_{ab}, u_{bc}, and u_{ca} and the output currents remain unchanged. As a matter of fact, the output currents i_a, i_b, and i_c depend on the line voltages. The sum of the three line voltages is equal to zero, $u_{ab} + u_{bc} + u_{ca} = 0$. Therefore, the set of three voltages comprises only two independent variables.

A more convenient representation of the output voltages is obtained by applying Clarke's transformation (2.9), which turns the phase voltages (u_a, u_b, u_c) into α, β, and the zero components. The zero component u_0 is the voltage of the floating star connection (u_{ST}), while the voltage components α and β define the voltage vector in an orthogonal α-β coordinate frame. In Fig. 2.14, the voltage vector is represented by its u_α and u_β components. The coil symbols represent the orientation of

Fig. 2.14 Clarke's
transformation represents
the phase voltages u_a, u_b,
and u_c as a voltage vector in
orthogonal α-β frame

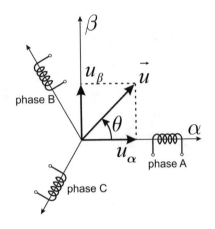

Table 2.1 Switching states and voltage vectors

Vector	SW code	u_a	u_b	u_c	u_α	u_β	$|u_{\alpha\beta}|$	Angle
V_0	000	$-E/2$	$-E/2$	$-E/2$	0	0	0	–
V_1	100	$+E/2$	$-E/2$	$-E/2$	E	0	E	0°
V_2	110	$+E/2$	$+E/2$	$-E/2$	$+E/2$	$+E\cdot\mathrm{sqrt}(3)/2$	E	60°
V_3	010	$-E/2$	$+E/2$	$-E/2$	$-E/2$	$+E\cdot\mathrm{sqrt}(3)/2$	E	120°
V_4	011	$-E/2$	$+E/2$	$+E/2$	$-E$	0	E	180°
V_5	001	$-E/2$	$-E/2$	$+E/2$	$-E/2$	$-E\cdot\mathrm{sqrt}(3)/2$	E	240°
V_6	101	$+E/2$	$-E/2$	$+E/2$	$+E/2$	$-E\cdot\mathrm{sqrt}(3)/2$	E	300°
V_7	111	$+E/2$	$+E/2$	$+E/2$	0	0	0	–

corresponding phase voltages, displaced by $2\pi/3$ and $4\pi/3$ with respect to the
phase A, which is aligned with abscise α.

$$\begin{bmatrix} u_\alpha \\ u_\beta \\ u_0 \end{bmatrix} = K \begin{bmatrix} 1 & -1/2 & -1/2 \\ 0 & +\sqrt{3}/2 & -\sqrt{3}/2 \\ 1/3 & 1/3 & 1/3 \end{bmatrix} \cdot \begin{bmatrix} u_a \\ u_b \\ u_c \end{bmatrix} (K=1) \qquad (2.9)$$

$$u_{ab} = u_\alpha - \frac{u_\beta}{\sqrt{3}}, u_{bc} = \frac{2u_\beta}{\sqrt{3}}, u_{ca} = -u_\alpha - \frac{u_\beta}{\sqrt{3}}. \qquad (2.10)$$

With $u_a + u_b + u_c = 0$, the line voltages u_{ab}, u_{bc}, and u_{ca} can be calculated from the
voltage components u_α and u_β (2.10). Under the circumstances, only the line
voltages affect the output currents. At the same time, the line voltages are uniquely
defined by u_α and u_β (2.10). Therefore, the voltage components u_α and u_β can be
used as a complete and compact representation of the output voltage.

The switching states of the three-phase inverter are listed in Table 2.1. For each
switching state, the table comprises the code, the phase voltages, the voltage
components u_α and u_β, the voltage amplitude, and the orientation of the voltage
vector in α-β coordinate frame. Six out of eight voltage vectors have a non-zero
amplitude, while the remaining two are zero.

Fig. 2.15 Position of eight
voltage vectors in
orthogonal α-β frame

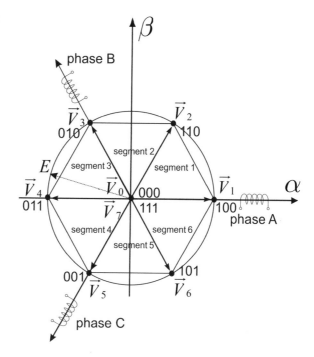

The voltage vectors listed in Table 2.1 are shown in Fig. 2.15. In orthogonal α-β frame, the vectors V_1–V_6 are located at the vertices of the hexagon. The hexagon is inscribed in the circle which has the radius E. The vectors V_0 and V_7 are located in the origin. Thus, the eight switching states of Fig. 2.13 are represented by seven discrete voltage vectors in α-β frame, each one denoted by a dot in Fig. 2.15.

2.2.2   The Switching Sequence and Dwell Times

Desired output voltage is represented by the voltage reference vector with an arbitrary amplitude and angle (Fig. 2.14). In most cases, the voltage reference does not coincide with any of the seven available voltage vectors. In order to obtain the output voltage with an average value equal to the reference, it is necessary to use several discrete voltage vectors within each PWM period.

In Fig. 2.16, it is assumed that the voltage reference resides within the first segment of the hexagon. Assuming that the switching pattern comprises the vectors V_1 and V_2, where the former dwells t_1, while the latter covers the rest of the switching period ($t_2 = T_{PWM} - t_1$), the average voltage within the switching period becomes

$$\vec{V}_{av2} = \frac{t_1\, \vec{V}_1 + t_2\, \vec{V}_2}{T_{PWM}}. \tag{2.11}$$

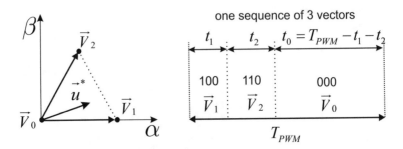

Fig. 2.16 The sequence of three voltage vectors

The average value calculated in (2.11) becomes equal to V_1 for $t_1 = T_{PWM}$ while reaching V_2 for $t_2 = T_{PWM}$. For $0 < t_1 < T_{PWM}$, it slides along the dashed line which connects the tips of the two vectors in Fig. 2.16. Thus, the sequence of two vectors would be sufficient for the voltage references that reside on the dashed line (Fig. 2.16), but it cannot provide the average voltage vector within the inner parts of segment 1.

The sequence of three voltage vectors is proposed in Fig. 2.16. The non-zero vectors V_1 and V_2 are maintained during the intervals t_1 and t_2, while the zero voltage vector V_0 dwells for the rest of the PWM period, $t_0 = T_{PWM} - t_1 - t_2$. With $V_0 = 0$, the average voltage of the three-vector sequence is given in (2.12), where $(t_1 + t_2)/T_{PWM} \leq 1$. The tip of the vector V_{av3} can be placed anywhere within the segment 1 by altering the dwell times t_1 and t_2.

$$\vec{V}_{av3} = t_1 \vec{V}_1 + t_2 \vec{V}_2 + t_0 \frac{\vec{V}_0}{T_{PWM}} = t_1 \vec{V}_1 + t_2 \frac{\vec{V}_2}{T_{PWM}}. \qquad (2.12)$$

The values t_1 and t_2 of the three-vector sequence can be calculated from the coordinates V_α and V_β of the reference vector. Considering (2.12) and the vector components listed in Table 2.1,

$$t_1 \vec{V}_1 + t_2 \frac{\vec{V}_2}{T_{PWM}} = \vec{\alpha}_0 \cdot V_\alpha + \vec{\beta}_0 \cdot V_\beta = \vec{\alpha}_0 \cdot \left(\frac{t_1}{T_{PWM}} E + \frac{t_2}{T_{PWM}} \frac{E}{2} \right) + \vec{\beta}_0 \cdot \left(\frac{t_2}{T_{PWM}} \frac{\sqrt{3}}{2} E \right). \qquad (2.13)$$

In (2.13), $\vec{\alpha}_0$ and $\vec{\beta}_0$ represent the unit vectors of the orthogonal coordinated system. The values t_1 and t_2 are obtained from

$$t_2 = \frac{2}{\sqrt{3}} \frac{V_\beta}{E} T_{PWM},$$

$$t_1 = \frac{V_\alpha}{E} T_{PWM} - \frac{t_2}{2} = \frac{V_\alpha}{E} T_{PWM} - \frac{1}{\sqrt{3}} \frac{V_\beta}{E} T_{PWM}. \qquad (2.14)$$

Generally speaking, the average voltage vector obtained by sequencing any three discrete vectors is a weighted sum of the discrete vectors, wherein the weight of each vector depends on the corresponding dwell time (2.12). The tip of the average voltage vector resides within the triangle defined by the tips of the three discrete vectors. Thus, whenever the reference voltage vector resides in segment 1, the three-vector sequence can be arranged with V_1-V_2-V_0. There are also other options; the segment 1 is contained by the triangle defined by the tips of the voltages V_1-V_2-V_5. Thus, it is possible to devise the three-vector sequence V_1-V_2-V_5 and to calculate the corresponding dwell times by using the approach outlined by (2.12, 2.13, and 2.14). Yet another three-vector sequence that meets the goal is V_1-V_2-V_4. For any of the aforesaid three-vector sequences, it is possible to calculate the dwell times that result in an average voltage vector that corresponds to the reference vector which resides in segment 1.

The choice of the three vectors affects the amplitude of the corresponding current ripple. The impact of the vector sequence can be envisaged from the equivalent circuit (Fig. 2.4, left) and the waveform (right) that represents the current ripple. The average value of the inverter voltage u_a in Fig. 2.4 corresponds to the back-electromotive force E_a. Therefore, any difference between the instantaneous and the average value of u_a is, at the same time, the difference between the instantaneous value of u_a and the back-electromotive force. In other words, the difference between the instantaneous and the average value of u_a is the voltage that appears across the series inductance L_a in Fig. 2.4. Larger voltages across the series inductance would cause an increase of the current ripple (2.2). Therefore, it is desirable to select the three-vector sequence where the voltage vectors remain close to the desired average voltage vector. In cases where the reference voltage resides in segment 1, obvious choices are V_1-V_2-V_0 and V_1-V_2-V_7. Other choices, such as V_1-V_2-V_5, would result in an elevated ripple due to the vector V_5 being at a considerable distance from the voltage reference. The selection of the three-vector sequences for the reference voltages that belong to segments 1–6 of the hexagon is given in Table 2.2. The sequences involve two non-zero voltage vectors, residing at the vertices of the hexagon, and one zero vector which resides at the origin. There are two zero vectors, V_0 and V_7. Thus, there are two three-vector sequences for each segment.

Table 2.2 Vector sequences for segments 1–6 of the hexagon

Segment	With zero vector V_0	With zero vector V_7
1	V_1-V_2-V_0	V_1-V_2-V_7
2	V_2-V_3-V_0	V_2-V_3-V_7
3	V_3-V_4-V_0	V_3-V_4-V_7
4	V_4-V_5-V_0	V_4-V_5-V_7
5	V_5-V_6-V_0	V_5-V_6-V_7
6	V_6-V_1-V_0	V_6-V_1-V_7

2.2.3 DD Switching Sequence

The sequence of voltage vectors defines the number of commutations within each PWM period. Each commutation is associated with energy losses. Therefore, the number of commutations per PWM period affects the efficiency of the inverter.

In three-phase inverters where the phases are controlled independently, there are two commutations in each period: the first when the lower switch (*AL* in Fig. 2.2) goes off while the upper switch (*AH*) goes on and the second when the upper switch goes off while the lower switch goes on. Thus, there is a total of six commutations per PWM period.

With space vector modulation, it is possible to reduce the number of commutations. The vector sequence in Fig. 2.17 involves three vectors, V_1, V_2, and the zero vector, which can be either V_0 or V_7. The sequence V_1-V_2-V_0 runs in counterclockwise direction. Repeating the pattern two times results in a "DD" sequence of the space vector modulation. The change from the switching state 100 to 110 requires one commutation. The change from 110 to 000 calls for two commutations, and the change from 000 back to the initial 100 calls for one more. Thus, the sequence of Fig. 2.17 requires four commutations in each period. The waveforms of the phase voltages which correspond to the DD switching sequence are given in Fig. 2.18.

In Fig. 2.18, the voltage u_c remains at $-E/2$. Namely, there are no commutations in phase C, as the lower switch *CL* remains on at all times. This reduces the commutation losses, but it also concentrates all of the conduction losses to the lower switch only. In cases where the DD sequence is organized with V_7, it is the phase A that would remain with the upper switch *AH* turned on at all times. The number of commutations can be further reduced by the DI sequence, described in the following subsection.

2.2.4 DI Switching Sequence

The DD switching sequence reduces the number of commutations per PWM period from six down to four. In each segment of the hexagon, one of the phases does not

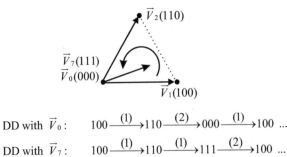

Fig. 2.17 Counterclockwise sequence of two non-zero vectors and one zero vector make up a "DD" sequence of space vector modulation

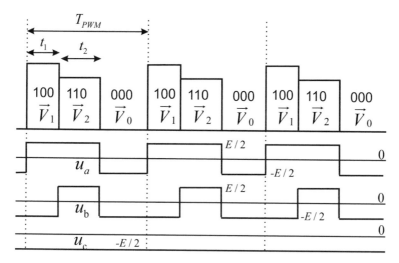

Fig. 2.18 DD vector sequences of the space vector modulation and the waveforms of corresponding phase voltages

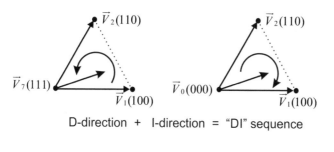

Fig. 2.19 The switching pattern with one counterclockwise sequence of voltage vectors followed by clockwise sequence makes up the "DI" sequence of the space vector modulation

have any commutation. The drawback of DD switching patterns is the fact that the all the commutation losses of non-commutating phase take place in one of the two switches.

The switching pattern can be organized by using one of the two zero vectors in odd PWM periods while using the other zero vector in even periods. If, at the same time, the patterns include turning from clockwise to counterclockwise sequences in adjacent periods, one obtains the "DI" sequence of the space vector modulation, illustrated in Fig. 2.19. The sequence uses both 000 and 111 zero vectors. At the same time, any transition between the two voltage vectors involves only one commutation. Compared to the DD sequence, the number of commutations is reduced from four to three.

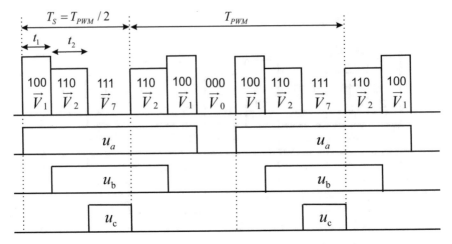

Fig. 2.20 DI vector sequences of the space vector modulation and the waveforms of corresponding phase voltages

The waveforms of the corresponding phase voltages are given in Fig. 2.20. Both 000 and 111 zero vectors are equally used, and the conduction losses are distributed between the upper and the lower switches. Considering the phase voltage, their period of change is two times larger than the sum of the dwell times in a three-vector sequence. That is, there are six discrete voltage vectors in each full period of the phase voltage waveforms. Thus, it is suitable to denote the length of the three-vector sequence by $T_S = T_{PWM}/2$ and to use T_{PWM} to designate the period of the phase voltages. The period T_S corresponds to the update rate $1/T_S$ in Fig. 2.8.

2.2.5 The Maximum Output Voltage with DI Sequence

With the three-vector sequence and with appropriate dwell times (2.14), the average value of the inverter output voltage can be placed anywhere within the hexagon of Fig. 2.15. In steady state, the inverter supplies the output voltage having a constant amplitude and frequency. In α-β orthogonal frame (Fig. 2.15), the output voltage is represented by the vector which revolves at a constant speed. The largest voltage amplitude is obtained at the vertices of the hexagon, and it is equal to $|u_{\alpha\beta}| = E$. Yet, it is not possible to supply the revolving vector of the amplitude E, since such vector would extend out of the hexagon. Therefore, the largest attainable voltage that can revolve within the hexagon maintaining the amplitude is equal to the radius of the inscribed circle:

$$\left| \vec{u}_{\alpha\beta} \right|_{max} = \frac{\sqrt{3}}{2} E. \tag{2.15}$$

From (2.10), the peak value of line-to-line voltages is calculated from the maximum voltage obtained in α-β frame:

$$u_{ab(\max)} = u_{bc(\max)} = \left|\vec{u}_{\alpha\beta}\right|_{\max} \cdot \frac{2}{\sqrt{3}} = E. \qquad (2.16)$$

Therefore, the maximum rms value of line voltages is equal to $U_{ab\ \mathrm{rms2}} = E/\mathrm{sqrt}$ $(2) = 0.707\ E$, the value which is $2/\mathrm{sqrt}(3) = 1.1547$ times larger than $U_{ab\ \mathrm{rms1}}$ in Question 2.3. The value $U_{ab\ \mathrm{rms1}}$ is calculated assuming that the phase voltages are sinusoidal, their peak value is $E/2$, and their rms value is $E/2/\mathrm{sqrt}(2)$, while the rms value of line voltages is $\mathrm{sqrt}(3)\ E/2/\mathrm{sqrt}(2) = 0.6124\ E$.

Since $U_{ab\ \mathrm{rms1}}$ represents the maximum rms value of the line voltages in cases where the phase voltages are sinusoidal, it is of interest to check the waveform of the phase voltages in cases where the voltages are larger, $U_{ab\ \mathrm{rms2}} > U_{ab\ \mathrm{rms1}}$. Considering the phase voltage waveforms in Fig. 2.20, it is observed that the voltage pulses of the amplitude $+E/2$ have the width of

$$t_{ON}^{A} = T_S + t_1 + t_2, t_{ON}^{B} = T_S + t_2 - t_1, t_{ON}^{C} = T_S - t_1 - t_2. \qquad (2.17)$$

For the first segment of the hexagon, the dwell times of individual vectors can be calculated from (2.14). Similar expressions apply for other segments. Consequently, the average value of the phase voltages u_a, u_b, and u_c can be found from (2.18).

The average value of the phase voltage and the line voltage obtained with DI sequence of the space vector modulation is plotted in Fig. 2.21. The plot represents the maximum voltages that can be achieved with the DI sequence of the space vector modulation. The same average value of the phase voltage can be obtained by using the scaled waveform of Fig. 2.21 as the modulation signal of the symmetrical PWM

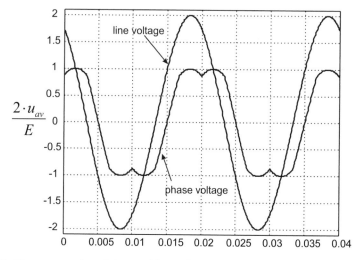

Fig. 2.21 The average value of phase and line-to-line voltages obtained with DI vector sequences of the space vector modulation

Fig. 2.22 The average value of the three-phase voltages and their sum obtained with DI vector sequences of the space vector modulation

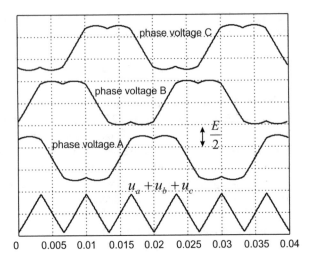

of Fig. 2.9. The phase voltage of Fig. 2.21 is not sinusoidal, but the difference between the two such phase voltages (u_a and u_b) results in a sinusoidal line-to-line voltage u_{ab} (line voltage in Fig. 2.21). The phase voltage u_a remains within the interval $[-E/2 .. +E/2]$, while the line voltage u_{ab} extends from $-E$ to $+E$.

$$u_{av}^{ABC} = \frac{2t_{ON}^{ABC} - T_S}{T_S} \cdot \frac{E}{2}. \tag{2.18}$$

The sum of the three-phase voltage is given in Fig. 2.22. As in the previous figure, the waveforms represent the average value of the corresponding voltages within each PWM period. The dwell times of the voltage vectors used within the first segment ($0 < \theta < \pi/3$) are given in (2.14). Similar expressions are used to calculate the dwell times in other segments. Thus, the analytical expressions that describe the phase voltages change between the segments. The same conclusion applies for the sum of the three-phase voltages. Using the expressions (2.14), (2.17), and (2.18) and focusing on the first segment ($0 < \theta < \pi/3$), the sum of the three-phase voltages is given by

$$(u_a + u_b + u_c)\big|_{0<\theta<\frac{\pi}{3}} = \left(\frac{E}{2}\right) \cdot \sqrt{3} \cdot \sin\left(\theta - \frac{\pi}{6}\right) \tag{2.19}$$

With the argument changing from $-\pi/6$ to $\pi/6$, the sine function is almost linear. With the analytical expression changes from one segment to another, the sum $u_a + u_b + u_c$ in Fig. 2.22 resembles a triangular waveform. Each of the phase voltages (u_a, u_b, u_c) in Fig. 2.22 has an amplitude of $E/2$, and it can be obtained by adding the signal ($u_a + u_b + u_c$)/3 to the sinusoidal waveform that has an amplitude of $2/\text{sqrt}$ (3) $E/2 \approx 1.15\ E/2 > E/2$.

Question (2.5) The three-phase inverter operates with DI sequence of the pulse width modulation and delivers sinusoidal line-to-line voltages of maximum amplitude. The average value of the phase voltages changes between -300 V and $+300$ V. What is the rms value of the line voltages?

Answer (2.5) The waveforms of the average phase voltage within each PWM period and the corresponding line-to-line voltage are shown in Fig. 2.21. The phase voltage waveform comprises the fundamental component and the triangular-looking waveform $(u_a + u_b + u_c)/3$ of Fig. 2.22. The amplitude of the fundamental component is 2/sqrt(3) times larger than 300 V. Thus, the amplitude of the line voltage is two times larger than $E/2$, while the rms value of the line voltage reaches 300 sqrt(2) $= 424.26$ V.

Question (2.6) The three-phase inverter feeds the three-phase R-L load and operates with DI sequence of the pulse width modulation and delivers sinusoidal line-to-line voltages of maximum amplitude. The dc-bus voltage is equal to 600 V. The voltage at the star connection of the load is not constant. What are the peak values of the star connection voltage?

Answer (2.6) The voltage at the star connection is equal to $(u_a + u_b + u_c)/3$. The waveform is shown in Fig. 2.22. The analytical expression that corresponds to the first segment of the hexagon is given in (2.19). The peak values are $\pm(E/2)$ sqrt (3) $\sin(\pi/6)/3 = \pm259.8/3 = \pm86.6$ V.

2.2.6   Symmetrical PWM with Common Mode Signals

The implementation of the space vector modulation is considerably more complex than the implementation of symmetrical PWM. The former starts with the segment selection and the selection of the two non-zero vectors, it proceeds by organizing the vector sequence and calculating the dwell times for each vector, and it ends by decoding the vector sequence into conventional PWM signals for each of the three phases. The latter requires just a plain comparison of the triangular PWM carrier and the modulation signal, but it provides lower output voltage than the space vector modulation. It is possible to calculate the modulation signals for the symmetrical, triangular carrier PWM that would provide the same waveforms of the phase voltages and line-to-line voltages as the DI sequence of the space vector PWM in Figs. 2.21 and 2.22.

Symmetrical PWM technique is illustrated in Figs. 2.7, 2.8, and 2.9. It comprises a triangular PWM carrier, and it generates the voltage pulses of the average value uniquely defined by the modulation signals. In cases where the frequency and amplitude of desired output voltages are constant, it is possible to introduce the modulation signal which comprises the sinusoidal waveform and the triangular-shaped correction $(u_a + u_b + u_c)/3$ shown in Fig. 2.22, described by (2.19) in the first segment $(0 < \theta < \pi/3)$ and by similar analytical expressions in other segments. In this case, and with symmetrical PWM, the average value of all the phase and line-to-line voltage within the subsequent PWM periods would correspond to the ones obtained from the space vector PWM (Figs. 2.21 and 2.22). Thus, any of the actual phase voltages would remain within $E/2$, yet the peak values of the corresponding fundamental components would reach beyond the $E/2$ limit, $2/\text{sqrt}(3)\, E/2 \approx 1.15\, E/2 > E/2$.

Triangular-shaped correction $(u_a + u_b + u_c)/3$ of (2.19) applies in cases where the desired voltages (and thus the modulation signals) correspond to constant-amplitude, constant-frequency sinusoidal waveforms. In general case, the output voltage may be different. One such example are the transients which may require non-sinusoidal change of the output voltages. The output voltages could be non-sinusoidal even in the steady state; an example is the grid-side inverter that injects sinusoidal currents into the ac grid with the voltages that comprise the line-frequency harmonics. For this reason, it is necessary to devise the signal that can substitute triangular-shaped correction $(u_a + u_b + u_c)/3$ of (2.19) in general case, even with non-sinusoidal modulation signals.

A closer insight into generation of the triangular-shaped correction $(u_a + u_b + u_c)/3$ can be obtained from Fig. 2.23. The three sinusoidal waveforms (m_a, m_b, m_c) represent the modulation signals. Compared to triangular PWM carrier of Fig. 2.9, the modulation signals that change from -1 to $+1$ produce the average value of the phase voltages that change from $-E/2$ to $+E/2$. The amplitude of the three sinusoids in Fig. 2.23 is $2/\text{sqrt}(3)$. The waveform marked by thick black lines is obtained by considering the waveforms m_a, m_b, and m_c and selecting the one with the smallest absolute value. Triangular-shaped correction $(u_a + u_b + u_c)/3$ of Fig. 2.22 and (2.19) can be replaced by taking one half of the waveform marked by the thick black lines in Fig. 2.23. This statement is supported by the following discussion.

In the first segment of the hexagon $(0 < \theta < \pi/3)$, triangular-shaped correction $(u_a + u_b + u_c)/3$ suggested by (2.19) is

$$\frac{(u_a + u_b + u_c)}{3} \cdot \frac{2}{E} = \frac{(m_a + m_b + m_c)}{3} = \frac{1}{\sqrt{3}} \cdot \sin\left(\theta - \frac{\pi}{6}\right) \qquad (2.20)$$

Fig. 2.23 Generation of the triangular-shaped correction of (2.19) and Fig. 2.22 from the three sinusoidal modulation signals

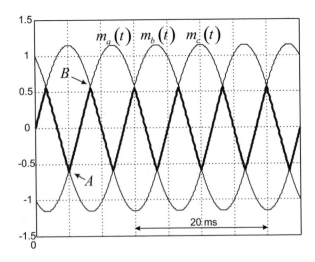

In Fig. 2.23, the interval marked by points A and B delimits a fragment of the function $(2/\sqrt{3})\sin(\theta - \pi/6)$ for $0<\theta<+\pi/3$. Thus, one half of the thick black waveform of Fig. 2.23 corresponds to the desired triangular-shaped correction of the phase modulation signals, shown in Fig. 2.22 and derived by (2.19) and (2.20).

Based on the previous discussion, desired common mode modulation signals $(m_a + m_b + m_c)/3$ can be obtained from the following procedure:

- The modulation signals $-1 < m_{a0} < +1$, $-1 < m_{b0} < +1$, and $-1 < m_{c0} < +1$ are obtained from the superior control levels (i.e., the current controller).
- At any given PWM period, it is necessary to select the modulation signal m_m with the smallest absolute value. If $|m_{a0}| < |m_{b0}|$ and $|m_{a0}| < |m_{c0}|$, then $m_m = m_{a0}$.
- The common mode modulation signal is obtained as $m_m/2$.
- The final modulation signals for the three phases are obtained by adding the common mode modulation signal to the original modulation signals, according to expressions $m_{a1} = m_{a0} + m_m/2$, $m_{b1} = m_{b0} + m_m/2$, and $m_{c1} = m_{c0} + m_m/2$.

The above procedure is written as a Matlab script in Table 2.3. For the purpose of demonstration, the procedure is applied to the case with sinusoidal output voltages. The procedure is delimited by the comments "Beginning" and "End." In a like manner, the procedure can be applied to an arbitrary set of non-sinusoidal modulation signals obtained at the output of the current controller or any other superior control layer. In any of such cases, it adds the common mode modulation signal to the phase quantities to enlarge the line-to-line voltages obtained at the output of three-phase inverters.

Table 2.3 Matlab script which calculates common mode modulation signal and modified modulation signals for the three phases

```
A = 2/sqrt(3);              % Assumption: Desired amplitude of the phase voltage
                            % is 1.15 * E/2, thus, it exceeds the DC bus voltage
for i = 1:720,              % Calculation covers two periods of the fundamental
  time(i) = i/720*0.04;
  anglerads = i*pi/180;
  ma0(i) = A*sin(anglerads);      % Sine modulation signals
  mb0(i) = A*sin(anglerads-2*pi/3);
  mc0(i) = A*sin(anglerads+2*pi/3);

% Beginning of the procedure >>>>>
  aa =abs(ma0(i)); bb = abs(mb0(i));cc = abs(mc0(i));% Finding the ABS values
  mmm = min(aa, bb);mmm = min(mmm, cc);             % Finding MIN(aa, bb, cc)
  if(mmm == aa)mm(i)= ma0(i);end;
  if(mmm == bb)mm(i)= mb0(i);end;                   % Selecting the smallest modulation
  if(mmm == cc)mm(i)= mc0(i);end;                   % signals
  ma1(i)= ma0(i)+ mm/2;
  mb1(i)= mb0(i)+ mm/2;                             % Adding the common mode
  mc1(i)= mc0(i)+ mm/2;
% End of the procedure        <<<<<

end;
plot(ma1; hold on; plot(ma1-mb1,'r');
```

Fig. 2.24 With
$u_{\text{GH}} = u_{\text{GL}} = -15$ V, both
transistor switches are off,
and the output current
$I_{\text{OUT}} > 0$ closes through the
lower diode D_L

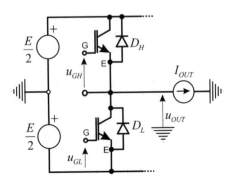

2.3   Lockout Time Error and Compensation

Ideally, the average value of the phase voltage u_{av} (2.1) is proportional to the modulation signal m (Fig. 2.9). The relation $u_{\text{av}} = m\, E/2$ is based on the assumption that the state of the power transistor switches (Fig. 2.24) is invariably defined by the PWM command and that the voltage drop across the conducting power switch is negligible. The PWM command is generated by the controller of Fig. 2.6 as a logic-level TTL signal. Practical implementation of one inverter phase requires the insertion of a brief, several microseconds wide interval between turning one of the switches off and turning the other one on. Such an interval is called *dead time* or *lockout time*. During this brief lockout time, the gate voltage across both of the switches is in off state (-15 V). This section discusses the need to use the lockout time, the negative consequences, and the available remedies.

2.3.1   Implementation of the Lockout Time

The n-phase voltage source inverter has n pairs of semiconductor power switches. The switches usually comprise a parallel connection of one power transistor and one power diode, where the former conducts positive and the latter negative currents. One inverter phase with one upper and one lower switch is shown in Fig. 2.24.

The power switches in Fig. 2.24 are controlled by the gate voltages u_{GH} and u_{GL}. For most common IGBT and MOSFET power switches, the gate voltage of $+15$ V brings the switch into conduction, while the gate voltage of -15 V keeps the switch off. The gate voltages are obtained from the gate drivers (Fig. 2.6) and controlled by the PWM command, obtained as a logic-level TTL signal from the digital controller.

In Fig. 2.25, the PWM command is equal to 1 before the instant t_1. At the same time, $u_{\text{GH}} = +15$ V and $u_{\text{GL}} = -15$ V. With $I_{\text{OUT}} > 0$, the upper power transistor in Fig. 2.25 is in conduction. Neglecting the voltage drop across the power transistor, $u_{\text{OUT}} = +E/2$ before t_1. At the instant t_1, the PWM command drops to zero, the gate voltage u_{GH} changes from $+15$ V to -15 V, and the upper power transistor turns off.

Fig. 2.25 The lockout time Δt_{DT} is the delay between the instant of turning off the preceding switch (t_1, t_3) and turning on the subsequent switch (t_2, t_4). The PWM command is obtained from Fig. 2.6 as logic-level TTL signal

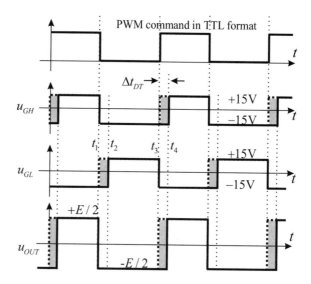

Regardless of the voltage u_{GL}, the current $I_{OUT} > 0$ commutates from the upper power transistor to the lower diode D_L, and the output voltage changes from $+E/2$ to $-E/2$.

It is of interest to notice that simultaneous conduction of both power transistors would create a destructive short circuit of the dc source E. When the upper transistor goes off, it is possible to turn on the lower power transistor without creating a short circuit by setting $u_{GL} = +15$ V.

During the interval shortly before t_1, an output current of the opposite sign ($I_{OUT} < 0$) turns into the upper diode D_H, as the power transistor cannot conduct a negative current, in spite of the positive gate voltage. In cases with $I_{OUT} < 0$, it is not necessary to keep $u_{GH} = +15$ V, since the current would keep the diode D_H in conduction even with $u_{GH} = -15$ V.

The gate voltage of $+15$ V is not required when the current turns into the parallel power diode, but it is compulsory for the opposite sign of the output current. The sign of I_{OUT} may not be readily available, as it may change in the course of the PWM period. Therefore, it is necessary to set $u_{GH} = +15$ V whenever the PWM command is high while keeping $u_{GL} = +15$ V whenever the PWM command is low (Fig. 2.25). An attempt to do so would require contemporary appearance of the rising edge of u_{GL} and the falling edge of u_{GH} (t_1).

The actual turning off of the power transistor starts with the gate voltage transition, and it ends when the transient phenomena within the semiconductor device terminate and the transistor regains its voltage-blocking capability. The transients complete in a very brief but finite interval of time. An attempt to turn on the complementary power transistor before the completion of the transients would lead to a brief but detrimental short circuit. The lockout time is inserted in order to prevent repetitive short-circuit spikes. At instant t_1 in Fig. 2.25, the gate voltage u_{GH} has transition from $+15$ V to -15 V, while the gate voltage u_{GL} remains at -15 V. After the lockout time delay Δt_{DT}, at instant t_2, any transient phenomena within the upper switch would come to an end, and the gate voltage u_{GL} turns to $+15$ V, turning the lower power transistor on. In a like manner, when the gate voltage u_{GL} has the falling edge at t_3, the gate voltage u_{GH} remains at -15 V for Δt_{DT}, before having a rising edge at t_4. Thus, there are two Δt_{DT}-wide intervals within each PWM period when both gate voltages are at -15 V.

2.3.2   The Voltage Error Caused by the Lockout Time

The switching transient decay time depends on the rated voltage and the type of the power transistor. The turn-off transients decay in 100 ns with low-voltage MOSFETs, but they could take more than 1 μs with high-voltage IGBTs. In most cases, the lockout time Δt_{DT} is set to considerably larger values, to account for variable delays in signal transmission lines, in optical transmitters, in receivers, and in gate driver filters and buffers. In Fig. 2.25, both gate voltages are at -15 V twice per period, during the intervals $[t_1 .. t_2]$ and $[t_3 .. t_4]$. With both power switches in blocking state, the current I_{OUT} turns into one of the diodes, D_L for $I_{OUT} > 0$, and D_H for $I_{OUT} < 0$.

The output voltage waveform (u_{OUT} in Fig. 2.25) is drawn for positive I_{OUT}. Within the interval $[t_1 .. t_2]$, the output voltage $u_{OUT} = -E/2$ corresponds to the PWM command, which is equal to zero, thus implying the desired voltage of $u^* = -E/2$. Within the interval $[t_3 .. t_4]$, the output voltage $u_{OUT} = -E/2$ does not correspond to the PWM command, which is equal to one, thus implying $u^* = +E/2$. Thus, the lockout time produces an error $\Delta u = u^* - u_{OUT} = E$ which extends over the interval $[t_3 .. t_4]$.

The voltage-error pulses are shown in Fig. 2.26. They have the width of Δt_{DT} and they repeat each T_{PWM}. The consequential error introduced into the system is

Fig. 2.26 The voltage error caused by the lockout time and its change with the sign of the output current I_{OUT}

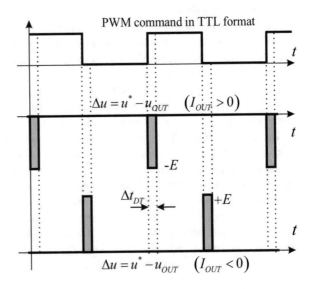

proportional to the ratio $\Delta t_{DT}/T_{PWM}$. For positive output current, the error pulses are negative. A negative output current produces positive error pulses. The occurrence of errors makes the output voltage different from the reference, and this may deteriorate the closed-loop performances of current controllers and other control loops.

2.3.3   Compensation of the Lockout Time Voltage Errors

The train of voltage-error pulses comprises PWM-related high-frequency components which do not affect the operation of the inverter loads that have a low-pass nature. At the same time, the error pulses have an average value. Within each PWM period, the average value of the voltage error is $E \, \Delta t_{DT}/T_{PWM}$. The sign of the error depends on the sign of the output current. In cases where the output current has a sinusoidal change (Fig. 2.27), the average value Δu_{av} of the voltage error becomes a square waveform. The sign of Δu_{av} is opposite to the sign of the output current. For $\Delta t_{DT} = 3$ µs and $T_{PWM} = 100$ µs, the amplitude of Δu_{av} reaches 3% of the dc-bus voltage, and such an error could introduce some detrimental errors and disturbances into the system.

In most cases, the output current is measured for the purposes of closing the feedback loops and for the protection. Therefore, it is possible to acquire the sign of all the three output currents. At the same time, the dc-bus voltage E is also measured. Thus, it is possible to calculate the necessary compensation for the modulation signals. The amplitude of such compensation is calculated from the ratio

Fig. 2.27 The average voltage error and its change with the sign of the output current I_{OUT}

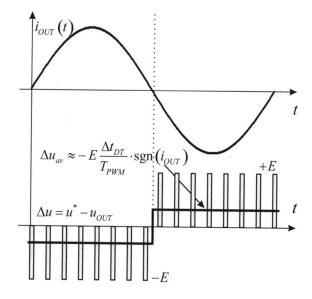

$\Delta t_{DT}/T_{PWM}$, while the sign is set according to the direction of the output currents. In the prescribed way, the error Δu_{av} of Fig. 2.27 can be accounted for.

Recommended compensation is based on the assumption that the output current turns to the power diodes during the lockout time interval. Rectangular form of the voltage error (Fig. 2.26) is based on such an assumption. The actual waveforms are somewhat different, as they include the power switch transients. Thus, instead of the rectangular waveforms of Fig. 2.26, the voltage waveform includes the rising slope of the voltage across the power switch during the turn-off transient. Such a waveform depends on the operating voltage, current, and temperature, and it is rather difficult to predict. Therefore, the suggested lockout time-error compensation reduces the voltage errors, but it does not eliminate them completely.

Question (2.7) The three-phase has the dc-bus voltage $E = 600$ V, the PWM period $T_{PWM} = 100$ μs, and the lockout time of $\Delta t_{DT} = 3$ μs. It has sinusoidal output currents that are in phase with the output voltages. The modulation signals are set to provide the largest sinusoidal output voltages, and they do not employ any lockout time compensation. Assuming that lockout time voltage error can be represented by its fundamental frequency component and that the voltage drop across the power switches is negligible, what is the rms value of the line voltages?

Answer (2.7) In absence of the lockout time, the rms value of the line voltages reaches E/sqrt(2) = 424.26 V. In each phase, the lockout time produces the voltage error which can be represented by the square-shaped waveform of the amplitude $E \, \Delta t_{DT}/T_{PWM} = 18$ V, synchronous with the output current and the output voltage. The first harmonic of the square-shaped waveform has the rms value of 4 18/π/sqrt (2) = 16.206 V. The line-to-line equivalent of that voltage is 28.07 V. Thus, the rms value of line-to-line voltages is 424.26–28.07 = 396.19 V.

Question (2.8) The inverter described in the previous question interfaces the three-phase mains through the three series inductances, $L = 5$ mH each. The modulation signals are set to provide sinusoidal line voltages, and they do not include the lockout time compensation. The mains voltages are sinusoidal, and they do not have any line-frequency harmonics. The rms value of the output current is close to the rated. What is the rms value of the fifth harmonic in the output current?

Answer (2.8) The first harmonic of the output voltages is in balance with the sum of the mains voltage and the first harmonic voltage drop across the series inductances. Since the mains voltage does not have any line-frequency harmonics, the fifth harmonic voltage drop across the series inductances is equal to the fifth harmonic of the inverter output voltages. As the modulation signals are set to provide sinusoidal voltages, the sole origin of the fifth harmonic is the lockout time voltage error. In each phase, the lockout time produces the voltage error represented by the square-shaped waveform of the amplitude $E \, \Delta t_{DT}/T_{PWM} = 18$ V. The fifth harmonic of such a waveform has an amplitude of (4/π) 18/5 = 4.5837 V and the rms value of 3.2411 V. The rms value of the consequential fifth harmonic in the output current is 3.2411/(L 5 ω) = 3.2411/(0.005 5 314.15) = 0.4127 A.

2.4 Design of the Output L Filters and LCL Filters

The output voltages of the PWM inverter comprise considerable high-frequency components, related to the switching frequency. The sample phase voltage waveform is given in Fig. 2.9 and its spectrum in Fig. 2.10. In most cases, the load has the low-pass nature due to ineluctable series inductance L. In cases where the inverter supplies an electrical machine, the low-pass nature comes from the equivalent series inductance of the windings. The sample waveform of the output current is given in Fig. 2.12. It comprises the fundamental component and the current ripple ΔI caused by the PWM voltages.

2.4.1 The rms Value of the Current Ripple

In a simplified single-phase representation (Fig. 2.4), the waveform of the current ripple is triangular. Its peak value is calculated in (2.2), $\Delta I_{max} = E \, T_{PWM}/(8 \, L)$. Taking into account the triangular form of the ripple, the rms value of the ripple waveform becomes

$$\Delta I_{rms0} = \frac{\Delta I_{max}}{\sqrt{3}} = \frac{E \cdot T_{PWM}}{8L\sqrt{3}}. \tag{2.21}$$

In both grid-side and electrical drive applications of three-phase inverters, the star connection of the equivalent load is left floating. Therefore, the phase current waveform gets affected by the voltage pulses in all the three phases. For this reason, the current ripple obtained with the floating star connection (Fig. 2.12) does not look like the waveform in Fig. 2.11, which has the triangular-shaped ripple.

The presence of the current ripple increases the rms value of the output current, but such an increase is rather low, as the ripple amplitude is just a fraction of the fundamental component. In PWM-supplied electrical machine, the current ripple at the PWM frequency may produce acoustical noise and contribute to additional losses in the windings and magnetic circuits of the machine. When injected into the ac grid, the current ripple may disturb the operation of other sources and loads that are connected to the same grid. Therefore, it is necessary to provide the output filter that attenuates the current ripple to an acceptable level. An L-filter has three series-connected inductances, one in each phase. The rms value of the ripple current obtained with an L-filter is given in Fig. 2.28. The ripple changes with the amplitude of the output voltage, given on abscise and represented by the maximum value of the modulation signals. The plot presents the rms value of the ripple divided by ΔI_{rms0} of (2.21).

Considering the fact that the grid-side inverters connect to ac grids with rather stable voltages, it is reasonable to assume that the peak values of the modulation signals change between 0.7 and 1. Therefore, the expression (2.21) provides an

Fig. 2.28 The change of the
relative rms value of the
current ripple with the
amplitude of modulation
signals

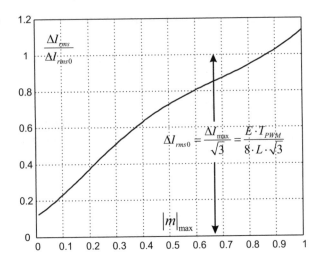

approximation of the current ripple with an error lower than 15%, quite suitable for
the purposes of the output filter design.

Attenuation of the *L*-type output filter is 20 dB per decade. In some cases, the
required attenuation can be obtained only with an exceptionally large series inductance, which introduces the problems of the size, weight, and the line-frequency
voltage drop. In such cases, it is necessary use an *LCL* filter with two series
inductances and one parallel capacitor in each phase.

2.4.2   The L-Type Output Filter

In most electrical drives, the electrical machine is fed from the three-phase PWM
inverter, shown in Fig. 2.2, where the windings of the electrical machine are
represented by symmetrical, star-connected load with series *R-L* impedance and
the back-electromotive force. It is the winding inductance *L* that attenuates the
PWM frequency component in the pulsed supply voltage and determines the rms
value of the current ripple (2.2). In cases where the winding inductance *L* is not
sufficient, a set of three external inductances can be connected in series.

The impact of the current ripple on the overall rms value of the output current is
usually very low. In cases where the peak value of the current ripple reaches 40% of
the rated current I_n, the rms value of the ripple is roughly 23.1% of I_n. The ripple
increases the overall rms value of the output current from 100% up to sqrt(100^2
$+ 23.1^2$) = 102.64%, contributing to an increase of only 2.64%. In electrical drives,
the ripple also contributes to the acoustic noise, to the torque ripple, and to additional
losses in the windings and magnetic circuits. In most cases, it is necessary to keep
I_{rms0} below 5%. From (2.21), this is achieved with an inductance:

$$L < \frac{E \cdot T_{\text{PWM}}}{8 \cdot \alpha \cdot I_n \cdot \sqrt{3}}, \tag{2.22}$$

where I_n is the rated output current, while $\alpha = 0.05$. The inductance value expressed in millihenries does not bring an immediate information on its size. The inductance size is better apprehended from the ratio $L \cdot \omega \cdot I_n / U_n$, which represents the relative voltage drop caused by the rated current I_n at the line-frequency ω_F. The rated rms value of the phase voltage is obtained from the dc-bus voltage E as $U_n = E/\text{sqrt}(3)/\text{sqrt}(2)$. In consequence,

$$L \frac{\omega_F I_n}{U_n} = \left(\frac{E \cdot T_{\text{PWM}}}{8\alpha I_n \sqrt{3}} \right) \frac{\omega_F I_n}{U_n} = \left(\frac{U_n \sqrt{6} T_{\text{PWM}}}{8\alpha I_n \sqrt{3}} \right) \frac{\omega_F I_n}{U_n} = \frac{\sqrt{2}\omega_F T_{\text{PWM}}}{8\alpha}. \tag{2.23}$$

For the line frequency of $\omega_F = 2 \cdot \pi \cdot 50$, the PWM period of $T_{\text{PWM}} = 100$ μs, and for the permissible rms value of the ripple current of 5% ($\alpha = 0.05$), the relative value of the required series inductance is 11.1%; that is, the rated current produces the voltage drop across the inductance equal to 11.1% of the rated voltage. This voltage drop is acceptable in many cases.

For grid-connected inverters, it is necessary to reduce the current ripple to considerably lower levels. In some cases (according to regulation IEEE 519), individual PWM-related spectral components of the output current should not exceed 0.3% of the rated current. According to results in Fig. 2.10, the most pronounced spectral component is the one at $f_{\text{PWM}} = 1/T_{\text{PWM}}$. By introducing $\alpha = 0.003$ into (2.23), the required series inductance should have the voltage drop in excess of 180% of the rated voltage, which is not feasible. In such cases, it is necessary to use an *LCL* output filter.

2.4.3   The LCL-Type Output Filter

The grid-side inverter with an *LCL* output filter is shown in the left side of Fig. 2.1. The objective of the subsequent analysis is calculation of the PWM-related ripple that passes from the inverter through the *LCL* filter into the grid.

The phase voltage spectrum (Fig. 2.10) comprises several PWM-related components. The most pronounced are the ones grouped next to the switching frequency f_{PWM}. Due to the low-pass nature of the filter, the voltage component at f_{PWM} causes the output current response which is two times larger than the same voltage at $2 \cdot f_{\text{PWM}}$. In consequence, the voltage components next to f_{PWM} have a prevailing impact on the rms value of the current ripple, and the design of the *LCL* filter can be performed by considering the attenuation at the switching frequency f_{PWM}, thus disregarding the spectral content at $2 \cdot f_{\text{PWM}}$, $3 \cdot f_{\text{PWM}}$, and other integer multiples of the switching frequency.

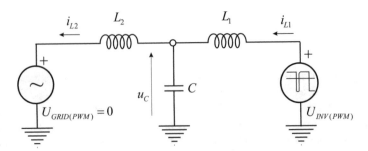

Fig. 2.29 LCL filter that suppresses the injection of PWM-related ripple from the grid-side inverter (on the right) into the ac grid (on the left)

When considering the voltage and current components at the switching frequency, the grid, the *LCL* filter, and the grid-side inverter of Fig. 2.1 can be represented by simplified circuit in Fig. 2.29. The grid-side inverter generates pulsed voltages which comprise significant component at the switching frequency ($U_{INV(PWM)}$). At the same time, it is assumed that the grid voltage does not have any component at the switching frequency, therefore, $U_{GRID(PWM)} = 0$. The voltage u_C and the currents i_{L1} and i_{L2} represent the frequency components at the PWM frequency.

The impedance $Z_C = 1/(2 \pi f_{PWM} C)$ of the parallel capacitor at the PWM frequency is rather low, as well as the voltage u_C. Thus, the ripple current i_{L1} at the inverter side can be calculated assuming that $u_C = 0$. The rms value of i_{L1} is limited by L_1. From (2.21), it is approximated by

$$I_{L1}^{rms} \cong \frac{E \cdot T_{PWM}}{8L_1 \sqrt{3}}. \tag{2.24}$$

One part of the current i_{L1} turns into the parallel capacitor, while the remaining part i_{L2} gets injected into the grid. The main objective of the *LCL* filter is reduction of the ripple injection i_{L2}. Focusing the analysis to the PWM frequency components, the rms value of the current i_{L2} is

$$I_{L2}^{rms} = \frac{I_{L1}^{rms}}{1 - L_2 \cdot C \cdot \omega_{PWM}^2} = \frac{\pi \cdot E}{4\sqrt{3}L_1\omega_{PWM}\left(1 - L_2 \cdot C \cdot \omega_{PWM}^2\right)}. \tag{2.25}$$

Design of the *LCL* filter consists in selecting the values of L_1, C, and L_2. The main objectives are reduction of the ripple current i_{L1} on the inverter side and the reduction of the ripple current i_{L2} injected into the grid.

Although the ripple i_{L1} has only a minor impact on the rms value of the inverter current, it increases the peak current in semiconductor power switches. Therefore, it is advisable to keep the peak value (2.2) of the current ripple below $0.3 I_n$ (30% of the rated current), which results in the rms value (2.24) of the ripple i_{L1} of I_n sqrt(3)/ 10 (roughly 17%).

Permissible injection of harmonic currents into the grid is rather low. According to the regulation IEEE-519, harmonics in the range of several kHz should not exceed 0.3% of the rated current. Thus, the rms value of the ripple current i_{L2} should not exceed I_n 0.003.

In addition to the above main objectives, there are also concerns related to the size of *LCL* components and undesirable resonances. Parallel connected capacitor draws certain line-frequency current and contributes to an increase in reactive power and an increase in rms value of the inverter current. Desirable reactive power in parallel capacitors is 5–10% of the rated power. Series-connected inductances contribute to the voltage drop at the line-frequency ω_F. The voltage drop across each of the inductances should not exceed 5% of the rated voltage, making the overall voltage drop lower than 10%. After all, the *LCL* filter should not enter into resonance. The filter gets connected to voltages which comprise the line-frequency components and the switching frequency components. Neither of these excitations should bring the *LCL* filter into resonance. Therefore, the resonant frequency of the filter should be far from either of the excitation frequencies. The above considerations could be expressed in terms of the following constraints:

$$Q_C = C \cdot \omega_F \cdot U_n^2 < \frac{U_n \cdot I_n}{20} \Rightarrow C < \frac{I_n}{20 \cdot \omega_F \cdot U_n}. \qquad (2.26)$$

$$(L_1 + L_2) \cdot \omega_F \cdot I_n < \frac{U_n}{10} \Rightarrow L_1 + L_2 < \frac{U_n}{10 \cdot \omega_F \cdot I_n}. \qquad (2.27)$$

$$\omega_{LCL} = \sqrt{\frac{L_1 + L_2}{L_1 L_2 C}}, \omega_F \ll \omega_{LCL} \ll 2\pi f_{PWM} = \omega_{PWM}. \qquad (2.28)$$

Design of the *LCL* filter results in different values for L_1 and L_2. For practical reasons, it is convenient to select $L_1 = L_2$ whenever possible.

Question (2.9) The three-phase grid-side inverter has the phase voltages of $U_n = 220$ Vrms and the rated line currents of $I_n = 10$ Arms. The switching frequency is 10 kHz and the dc-bus voltage is $E = 600$ V. It is necessary to design the *LCL* filter using the expressions (2.24, 2.25), respecting the constraints (2.26, 2.27, and 2.28), and abstaining from the requirement $L_1 = L_2$.

Answer (2.9) Assuming that the PWM component of the voltage u_C is negligible, the current i_{L1} is limited by the inductance L_1. Assuming that the rms value (2.24) of the ripple i_{L1} is limited by I_n sqrt(3)/10, the inductance L_1 is

$$L_1 \geq \frac{10 \cdot E \cdot T_{PWM}}{8 \cdot 3 \cdot I_n} = 2.5\text{mH}$$

With $L_1 = 2.5$ mH, the rms value of the current ripple i_{L1} is 1.7321 A. Permissible value of the rms value of the current ripple i_{L2} is 0.003 I_n. With the resonant frequency of the *LCL* filter considerably lower than the switching frequency, it is reasonable to assume that $L_2 C \omega^2_{\text{PWM}} \gg 1$. Therefore, attenuation expressed in (2.25) can be simplified to

$$I^{\text{rms}}_{L2} \approx \frac{I^{\text{rms}}_{L1}}{L_2 \cdot C \cdot \omega^2_{\text{PWM}}}.$$

In order to achieve the required attenuation, it is necessary to select L_2 and C so as to obtain $L_2 C \omega^2_{\text{PWM}} > 57.74$. For the given switching frequency, desired product $L_2 C$ has to reach 1.4626e-08.

From (2.26), the upper limit for the parallel capacitor is $C_{\text{max}} = 7.235$ µF. From (2.27), the upper limit for the inductances is $(L_1 + L_2)_{\text{max}} = 7$ mH. The first inductance is already set to 2.5 mH, leaving the space for $L_{2\text{max}} = 4.5$ mH. Selecting the maximum permissible values for L_2 and C, the factor $L_2 C \omega^2_{\text{PWM}}$ reaches 128.5, far beyond the required 57.74. Thus, there is an infinite number of L_2 and C values that provide the necessary $L_2 C$ product. This degree of freedom can be used to meet yet another design goal.

Frequently desired property of L, LC, and LCL filters is reduced size of the associated capacitors and inductances. In cases where the LC product is already set, selection of a small inductance would require a large capacitor and vice versa. The solution could be found by selecting the values of L and C in the way that provides the same reactive power in both components, namely, $L \omega_{\text{PWM}} I_n^2 = C \omega_{\text{PWM}} U_n^2$. From the later condition and from the requirement $L_2 C = 1.4626\text{e-}08$,

$$Q_{L2} = Q_C \Rightarrow \frac{L}{C} = \left(\frac{U_n}{I_n}\right)^2 = 484\Omega^2, L_2 \cdot C = 1.4626 \cdot 10^{-8},$$
$$\Rightarrow C = 5.5\mu F, L_2 = 2.66\text{mH}.$$

This completes the design with $L_1 = 2.5$ mH, $L_2 = 2.66$ mH, and $C = 5.5$ µF.

Question (2.10) The grid-side inverter with the same rated voltages, the same currents, and the same dc-bus voltage as in the previous question has the switching frequency that is 16 kHz. It is necessary to design the LCL filter assuming that $L_1 = L_2$ and that the reactive power is the same in each of the three components $(Q_{L1} = Q_C = Q_{L2})$.

Answer (2.10) Under assumptions outlined in the previous question, the expression (2.25) can be simplified to

$$I^{\text{rms}}_{L2} = 30\text{mA} \approx \frac{\pi \cdot E}{4\sqrt{3}L_1 \cdot L_2 \cdot C \cdot \omega^3_{\text{PWM}}}.$$

Assuming that $L_1 = L_2 = L$,

$$L^2 \cdot C = \frac{\pi \cdot E}{4\sqrt{3} \cdot 0.03 \cdot \omega^3_{\text{PWM}}} = 8.926 \cdot 10^{-12}.$$

It is required that, at rated operating conditions, each of the series inductances has the same reactive power as the parallel capacitor. In the previous question, it has been shown that such a requirement results in the relation:

$$\frac{L}{C} = \left(\frac{U_n}{I_n}\right)^2 \Rightarrow L = 484 \cdot C.$$

From the previous two expressions, $C = 3.365\ \mu F$ and $L = L_1 = L_2 = 1.629$ mH. Reactive power on each of the series inductances is equal to 2.326% of the rated power, the same as the reactive power across the parallel capacitor. The voltage drop across the two series-connected inductances is 4.652% of the rated voltage. The rms value of the inverter current in (2.24) is 1.66 A, which is lower than 17% of the rated current. The rms value of the current injected into the grid (2.25) is 30 mA, as requested.

2.5   Multilevel Inverters and Their PWM Techniques

The inverters illustrated in Figs. 2.2 and 2.6 have two discrete voltage levels in each phase, $-E/2$ and $+E/2$. It is for this reasons that they are called the two-level inverters. The two-level inverters have several advantages:

- Two-level inverters have a relatively low number of power diodes and power transistors, two diodes and two power transistors in each phase.
- A low number of power switches and a low number of gate drivers keep the auxiliary electronics rather simple, as well as the generation of the TTL command signals.
- In each of the phases, only one of these devices is in conduction at each instant, thus resulting in relatively low conduction losses.

At the same time, the two-level inverters also have some drawbacks:

- When in blocking state, each of the devices must sustain full dc-bus voltage E. Thus, the breakdown voltage of power devices must be considerably larger than E.
- At each commutation, the corresponding phase voltage changes by E in a fraction of microsecond (50 ns–100 ns), contributing to a very large dV/dt stress that has adverse effects on semiconductor power devices, insulation materials, and the windings.
- A large amplitude of the PWM voltage pulses contributes to a large current ripple. In turn, suppression of the ripple requires larger L-C elements of the output LCL filter.

2.5.1 Three-Level Inverters

The problems associated with two-level inverters can be resolved or alleviated by selecting the inverter topology which outputs the voltage that can assume more than two discrete levels. One phase of the three-level inverter is shown in Fig. 30. Depending on the status of the power switches, the output voltage can assume one of the three discrete values, $-E/2$, 0 or $+E/2$.

There are four semiconductor power switches in each phase. When the TTL command signals are $S1 = S2 = 1$ and $S3 = S4 = 0$, both Q_1 and Q_2 are turned on, and the output voltage is equal to $+E/2$. The switches Q_3 and Q_4 are in off state. The voltage across the series connection of these switches is equal to E. If the voltage is equally shared between the switches Q_3 and Q_4, each of them has to sustain the blocking voltage of $E/2$. Thus, the voltage rating of semiconductor power switches in a three-level inverter is one half of the voltage rating required in a two-level inverter. While the output voltage is $u_a = +E/2$, the output current flows through both Q_1 and Q_2, contributing to the voltage drop and to the conduction losses in both switches. When the output current is negative and the output voltage is $u_a = +E/2$, the current flows through D_1 and D_2.

With the TTL command signals set to $S1 = S2 = 0$ and $S3 = S4 = 1$, the two lower switches Q_3 and Q_4 are turned on, and the output voltage is equal to $-E/2$. The switches Q_1 and Q_2 are turned off, and each of them sustains the blocking voltage of $E/2$. For negative output currents, the output current flows through the series-connected Q_3 and Q_4. Positive output currents pass through the series-connected diodes D_3 and D_4.

In cases where the TTL command signals are set to $S1 = S4 = 0$ and $S2 = S3 = 1$, the switches Q_1 and Q_4 are turned off, while the switches Q_2 and Q_3 are turned on. Positive output currents flow through the series connection of D_5 and Q_2. Negative output currents flow through the series connection of D_6 and Q_3. Neglecting the voltage drop across the conducting power switches, the output voltage u_a is equal to zero in both cases. All the output voltages are referred to the middle point of the dc-bus, namely, the node which connects C_1 and C_2 in Fig. 2.30.

In three-level inverters, the middle point of the dc-bus is also called the *neutral point*. Activation of the power switches Q_2 and Q_3 connects the output voltage to the *neutral point*, whatever the direction of the output current. In a way, the output phase gets *clamped* to the neutral point. When *clamped*, the output voltage is equal to zero. For this reason, the three-level topology in Fig. 2.30 is called *neutral point clamped* or *NPC*. There are also other converter topologies that provide multiple levels of the output voltage, and one of them is the *flying capacitor* circuit.

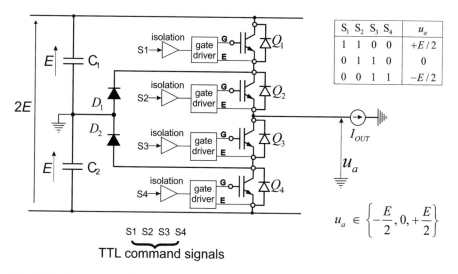

S_1	S_2	S_3	S_4	u_a
1	1	0	0	$+E/2$
0	1	1	0	0
0	0	1	1	$-E/2$

$$u_a \in \left\{ -\frac{E}{2}, 0, +\frac{E}{2} \right\}$$

Fig. 2.30 One phase of the three-level inverter. It takes four power transistors, four TTL commands, and six diodes to generate $u_a(t)$ with three discrete levels

2.5.2   The Phase Voltages and Line Voltages

The three-phase three-level inverter with the star-connected load comprising the electromotive forces and series R-L impedances is shown in Fig. 2.31. Compared to the equivalent two-level inverter, the number of power transistors is doubled, and the number of power diodes is tripled. At the same time, the current flows through two series-connected devices at each instant, contributing to increased conduction losses. At the same time, each output voltage has two discrete levels, $-E/2$, 0 and $+E/2$. Compared to the two-level inverter, the blocking voltage across the power devices is halved, along with the dV/dt stress.

The waveforms in Fig. 2.32 represent the phase voltage and the line voltage of the three-level inverter. The fundamental frequency of the output voltages is 50 Hz. The phase voltage average value within each PWM period has a sinusoidal change over time.

The pulse width modulation applied in Fig. 2.32 is obtained by using the PWM technique illustrated in Fig. 2.9 and by introducing the changes that are required to meet the needs of the three-level inverter:

- Triangular-shaped PWM carrier $m_{PWM}(t)$ corresponds to the one in Fig. 2.9. While the later changes from -1 to $+1$, the signal $m_{PWM}(t)$ changes from 0 to $+1$.
- The modulation signal $m_a(t)$ is sinusoidal, and it corresponds to the modulation signal in Fig. 2.9. The frequency of $m_a(t)$ is set to 50 Hz. The signal changes between $-A$ and $+A$. The amplitude A changes from 0 to 1. For $A = 1$, the

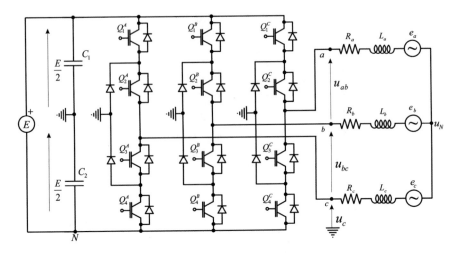

Fig. 2.31 Three-phase three-level inverter connected to the three-phase load

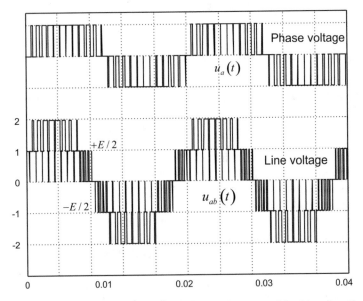

Fig. 2.32 The phase voltage and the line-to-line voltage at the output of the three-phase three-level inverter. The waveforms are obtained with modified triangular carrier-based modulation of Fig. 2.9

sinusoidal change of the one-PWM-period average of the phase voltage reaches the maximum value of $E/2$.

- The signals $0 < |m_a(t)| < 1$ and $0 < m_{PWM}(t) < 1$ are brought to the comparator which outputs the signal $S_{AA} = 1$ when $|m_a(t)| > m_{PWM}(t)$ and $S_{AA} = 0$ when $|m_a(t)| < m_{PWM}(t)$.

- The signal S_{SS} is created as the sign of the modulation signal $m_a(t)$. It is equal to $+1$ when $m_a(t) > 0$ and -1 otherwise.
- When $S_{SS} = 1$ and $S_{AA} = 1$, the signals $S_1{:}S_2{:}S_3{:}S_4$ (Fig. 2.30) are set to "1100," the switches Q_1 and Q_2 are turned on, and the output voltage is equal to $u_a = +E/2$.
- When $S_{SS} = -1$ and $S_{AA} = 1$, the signals $S_1{:}S_2{:}S_3{:}S_4$ are set to "0011," the switches Q_3 and Q_4 are on, and the output voltage is $u_a = -E/2$.
- When $S_{AA} = 0$, the signals $S_1{:}S_2{:}S_3{:}S_4$ are set to "0110," the switches Q_2 and Q_3 are on, and the output voltage is $u_a = 0$.

Corresponding waveforms in Fig. 2.32 are obtained $A = 1$. The phase voltage has three discrete levels, while the line-to-line voltage has five discrete levels. For this reason, the current ripple caused by the non-sinusoidal supply is considerably lower than with the two-level inverters. The same approach can be used to design the inverters with even more discrete levels of the output voltage. By increasing the number of voltage levels, the output voltage waveforms are brought closer to the sinusoidal shape. Multilevel inverters are often used in applications with large dc-bus voltage, as the voltage rating of their power switches is lower than the bus voltage. They are also used in cases where the PWM frequency is relatively low, since the multilevel output waveforms reduce the current ripple.

2.5.3 Space Vector Modulation in Multilevel Inverters

The space vector modulation concept is described in Sect. 2.2.2, where it is applied in two-level inverters. Instead of an independent control in individual phases, the concept proposes simultaneous control which takes into account the commutations in all the phases. The space vector modulation includes the selection of the convenient output vectors and the design of the three-vector sequences that would provide the desired output with reduced number of commutations.

The same concept can be applied to multilevel inverters. For simplicity, the switching state in each of the phases can be described by one digit equal to 0, 1, or 2. The values 0, 1, and 2 correspond to the output voltages $-E/2$, 0, and $+E/2$. Associating one such digit to each phase, it is possible to describe the switching state of the three-phase inverter by a simple three-digit switching code, such as 210. The first digit (2) implies $u_a = +E/2$, the second digit (1) $u_b = 0$, and the third digit (0) implies $u_c = -E/2$. The available switching states can be presented in three-dimensional a-b-c space, where each of the axes corresponds to one of the inverter phases.

Each switching code corresponds to one set of phase voltages (u_a, u_b, u_c). By using Clarke's transformation (2.9), the three voltages can be represented by one voltage vector in α-β coordinate frame (Fig. 2.14). Thus, each switching state corresponds to one of the available voltage vectors. Notice in Fig. 2.33 that one and the same voltage vector could be obtained from different switching states. An example is the zero vector $(u_\alpha = 0, u_\beta = 0)$, which can be obtained from the

Fig. 2.33 The switching states of the three-level three-phase inverter. Each switching state is described by a three-digit code. The digits 0, 1, and 2 correspond to the output voltages $-E/2$, 0, and $+E/2$. The first digit in each code corresponds to the phase a, the second to the phase b, and the third to the phase c

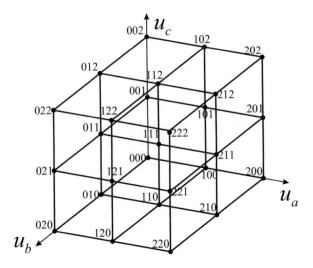

switching codes 000, 111, and 222. Another example is the vector ($u_\alpha = +E/2$, $u_\beta = 0$), which can be obtained by applying the switching code 100 but also the switching code 211.

The voltage vectors in Fig. 2.34 can be divided in three groups, the vectors on the outer hexagon (such as 200, 210, 220, etc.), the vectors on the inner hexagon (such as 100/211, 110/221, 010/121, etc.), and the zero vectors (000, 111, 222). There are three switching states that provide the zero vector, two switching states for each voltage vector on the inner hexagon, and only one switching state for the voltage vectors on the outer hexagon. Alternative switching states could be used to reduce the number of commutations within each vector sequence.

In Fig. 2.34, the reference voltage u^* resides within shaded triangle. In order to generate the average voltage which corresponds to the reference, it is necessary to design a sequence of three voltage vectors and to determine their dwell times in the manner similar to (2.12).

Like in two-level space vector modulation, the distance between the reference voltage u^* and the selected voltage vectors governs the current ripple. Therefore, it is convenient to select the three nearest voltage vectors.

Considering the voltage reference u^* given in Fig. 2.34, the three adjacent vectors are obtained with the switching codes 210, 221/110, and 211/100. In order to reduce the number of commutations down to four per cycle, it is convenient to select the codes 210, 221, and 211. Thus, the average output voltage u^* can be obtained by cycling the sequence 210:221:211 with corresponding dwell times obtained from

$$\frac{t_{210} \cdot \vec{V}_{210} + t_{221} \cdot \vec{V}_{221} + t_{211} \cdot \vec{V}_{211}}{t_{210} + t_{221} + t_{211}} = \vec{u}^* = \vec{\alpha}_0 \cdot u_\alpha^* + \vec{\beta}_0 \cdot u_\beta^*. \qquad (2.29)$$

Fig. 2.34 The output voltage vectors of the three-phase three-level inverter in α-β coordinate frame. The reference voltage vector resides within the shaded triangle

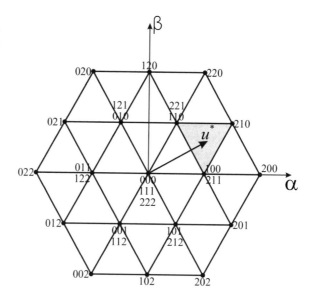

The sum of the three dwell times is equal to the cycle time, $T_{\mathrm{PWM}} = t_{210} + t_{221} + t_{211}$. The vectors α_0 and β_0 are the unit vectors of α-β coordinate frame. The output vectors in (2.29) can be expressed in terms of their α-β components:

$$\vec{V}_{210} = \vec{\alpha}_0 \cdot E\frac{3}{4} + \vec{\beta}_0 \cdot E\frac{\sqrt{3}}{4}, \ \vec{V}_{221} = \vec{\alpha}_0 \cdot \frac{E}{4} + \vec{\beta}_0 \cdot E\frac{\sqrt{3}}{4}, \ \vec{V}_{211}$$

$$= \vec{\alpha}_0 \cdot \frac{E}{2} + \vec{\beta}_0 \cdot 0. \tag{2.30}$$

The dwell times of the three vectors are obtained from (2.29):

$$\begin{aligned} t_{210} &= T_{\mathrm{PWM}} \cdot \left(\frac{2 \cdot u_\alpha}{E} + \frac{2 \cdot u_\beta}{E \cdot \sqrt{3}} - 1 \right), \\ t_{221} &= T_{\mathrm{PWM}} \cdot \frac{4 \cdot u_\beta}{E \cdot \sqrt{3}} - t_{210}, \\ t_{211} &= T_{\mathrm{PWM}} - t_{210} - t_{221}. \end{aligned} \tag{2.31}$$

In Fig. 2.34, there are 24 equilateral triangles with their sides equal to $E/2$. Whenever the voltage reference u^* befalls in one of these triangles, the three-voltage-vector sequence is arranged from the vectors that reside at vertices of the triangle. The dwell times are calculated from considerations similar to those in (2.29, 2.30, and 2.31).

2.6   Summary

In grid-side power converters and electrical drives, there is a need to provide the ac voltages with adjustable amplitude, frequency, and phase. In most cases, the system comprises an internal current loop, designed and set to bring the ac current to desired reference value. The voltage is used as the driving force within digital current controllers. The voltages are obtained from PWM inverters, which assume the role of the voltage actuators.

The inverters cannot provide continuous change of output voltages. Instead, they provide the output voltages that assume one of the discrete values. In most common two-level inverters, each phase voltage can be equal either to $-E/2$ or to $+E/2$. Desired output voltage u^* changes between $-E/2$ and $+E/2$. The instantaneous value of the voltage u_a at the output of the switching bridge cannot be brought to an intermediate value u^*. Instead, the voltage u_a is generated as a series of voltage pulses that repeat with the period of T_{PWM}. The average value of the pulsed output voltage is changed by adjusting the pulse width. In cases where the pulse width changes from one PWM period to the other in a sinusoidal manner, the average value of the output voltage becomes a sinusoidal function of time.

The change of the one-PWM-period average of the output voltage corresponds to the desired voltage reference. In addition to this average value, the pulsed output voltages also comprise the components at the PWM frequency and its multiples. This high-frequency content produces the current ripple. In order to suppress the ripple, the grid-side inverters are connected to the grid through the low-pass LCL filters. By putting aside the high-frequency contents of the PWM voltages and by compensating the lockout time imperfections, the PWM inverters could be treated as ideal voltage actuators.

Chapter 3
Acquisition of the Feedback Signals

Closed-loop performance of grid-side inverters and electrical drives depend on the closed-loop bandwidth of digital current controllers, which are used as inner control loop in vast number of cases. The current control is organized on the feedback principles. Thus, it is necessary to measure the output currents for the purpose of closing the feedback and also for the purpose of overcurrent protection. The feedback acquisition chain comprises analogue and digital devices and filters. Their purpose is producing the digital word that represents the output current with the least possible delay and error. The digital word resides within RAM of the digital controller, and it is used for the feedback and protection purposes.

The current sensors provide the voltage or current signals that duplicate the waveform of the output current. It is necessary to provide the galvanic insulation and pass the signals to the digital controller. Before the sampling, it is necessary to apply anti-alias analogue filters which suppress the sampling errors. Further on, equidistant samples are passed to the A/D converter which turns the analogue signals into digital words. The process gives rise to the lack of information through both time quantization (sampling) and amplitude quantization (A/D conversion). At the same time, the PWM switching noise contributes to the errors in the feedback chain. Design objectives include suppression of the PWM noise, minimization of the quantization errors, and reduction of the consequential delays.

3.1 Current Sensors and Galvanic Insulation

In twin-inverter setup of Fig. 2.1, both the grid-side inverter and the generator-side inverter require the output current measurement. In cases where the sum of the three output currents is zero ($i_a + i_b + i_c = 0$), it is possible to employ only the sensors for i_a and i_b, while the third current can be obtained from the remaining two ($i_c = -i_a - i_b$).

The current sensors generate signal-level voltages or currents that are proportional to the measured output current. The sensors can be plain shunt resistors,

© Springer International Publishing AG, part of Springer Nature 2018
S. N. Vukosavic, *Grid-Side Converters Control and Design*, Power Electronics and Power Systems, https://doi.org/10.1007/978-3-319-73278-7_3

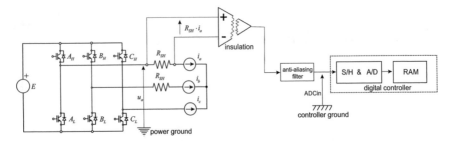

Fig. 3.1 Current measurement system with shunt resistors R_{SH}. Dangerous power circuit voltages have to be isolated from the digital controller. Thus, the analogue signal $R_{SH} \cdot i_a$ has to pass through isolation amplifiers

current transformers, Rogowski coils, Hall effect devices, fluxgate sensors, or fiber-optic current sensors with interferometer.

3.1.1 Shunt-Based Current Sensing

The use of shunt resistors is illustrated in Fig. 3.1. The voltage across the two shunt resistors is proportional to the output currents i_a and i_b. In systems where the star connection of the load is floating, and assuming that there are no ground faults, the third current can be calculated from the other two. Therefore, there are only two R_{SH} resistors in the figure.

The dc-bus voltage and the line voltages have to be isolated from the controller ground. Therefore, it is necessary to feed the signals $R_{SH} \cdot i_a$ and $R_{SH} \cdot i_b$ to the input of isolation amplifiers, devices capable of passing the analogue signals through galvanic isolation. The output of the isolation amplifier refers to the controller ground.

The PWM noise and other disturbances encountered in power electronics environment could impair the sampling process and contribute to errors in feedback signals. Therefore, the output of the isolation amplifier is passed to the analogue filter, designed to suppress the high-frequency components that could lead to erroneous sampling. Filtered signal (*ADCin*) is brought to the sample-and-hold circuit (S/H), which holds the acquired sample during the process of A/D conversion. Conversion result is stored into designated RAM and used as the current feedback.

The shunt resistor in the output phase A has the power dissipation of $R_{SH} \cdot i_a^2$. Assuming that the voltage drop $R_{SH} \cdot i_a$ across the shunt is 100 mV and that the output current is $i_a = 500$ A, the power dissipation of the shunt is 50 W. Corresponding value of the shunt is 0.5 mΩ. For practical reasons, it is of interest to keep the power dissipation low; thus, there are good reasons to seek for lower shunt resistances and lower voltage drop.

For the proper operation of the isolation amplifier, the magnitude of the input signal $R_{SH} \cdot i_a$ has to be at least 100 mV. This is required in order to maintain the signal-to-noise and the signal-to-offset ratios. Namely, the input section of isolation amplifiers is similar to the one of conventional operational amplifiers, and it has finite values of the offset, drift, and noise. Reduction of the useful signal below 100 mV would increase the relative impact of the offset and noise. In such case, one and the same offset between the input terminals of the amplifier would contribute to a larger measurement error, expressed in amperes of the output current.

There is also another problem associated with low-resistance shunts. For the shunt values as low as 0.5 mΩ, the voltage drop across the leads becomes significant. At the same time, the leads of the shunt resistor have a small but finite inductance L_{SH}. Thus, the voltage across the shunt is actually

$$u_{SH} = R_{SH} \cdot i_a + L_{SH} \frac{di_a}{dt}. \tag{3.1}$$

With the load current that has the ripple of 100 A at 10 kHz, the slope of the output current reaches $di_a/dt = 4$ A/μs. With parasitic inductance of the shunt leads of only $L_{SH} = 25$ nH, the voltage drop across L_{SH} reaches 100 mV, the level of the useful signal. For this reason, shunt resistors are made with four terminals. Two of them are used to feed the output current i_a to a low-inductance resistor, while the remaining two are just tiny signal lines that pick up the voltage across the low-inductance resistor. In order to reduce the internal inductance, the low-inductance resistor is usually made as a thick-film device.

Regardless of the need to employ isolation amplifiers, the shunt resistor current sensing is a low-cost solution, simple to design and use. Due to necessity to keep the voltage drop across the shunt resistor above 100 mV, the shunt sensing of the output currents is not convenient for large power inverters. The shunt sensing is mostly used in low-power moderate-performance electrical drives where the cost is of particular importance.

3.1.2   Current Transformers

Current transformers are the standard solution for the current sensing in power systems. It is rugged, low-cost sensor with very high precision and bandwidth. Assuming that the output current of the inverter is 500 A, the current transformer with the transformation ratio $N_2{:}N_1 = 5000{:}1$ (Fig. 3.2) provides the secondary current $i_2 = 100$ mA. The voltage across the shunt resistor $R_{SH} = 10\ \Omega$ is $u_{SH} = 1$ V. The voltage u_{SH} is passed through the anti-aliasing filter and brought to the A/D converter.

The shunt resistor in the considered sample circuit has the power dissipation of only 100 mW. Thus, the circuit of Fig. 3.2 could be operational even with the current transformer designed for the rated power considerably lower than $S_n = 1$ VA. Yet,

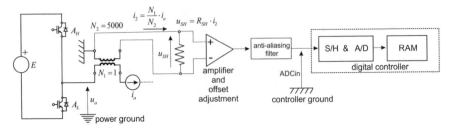

Fig. 3.2 Current measurement system with the current transformer. The primary and secondary circuits are isolated. The secondary current i_2 is N_2/N_1 times lower than the primary current i_a, and it provides the analogue signal $R_{SH} \cdot i_a$, proportional to the current being measured

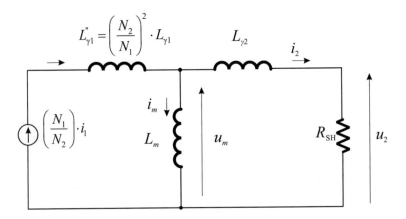

Fig. 3.3 Simplified equivalent circuit of the current transformer

the current transformers are designed with S_n considerably larger than $u_{SH} \cdot i_2$. During short-time overloads, the voltage u_{SH} can exceed the rated value by an order of magnitude before the magnetic circuit of the current transformer gets into magnetic saturation. In other words, the current transformers are designed for the operation with very low voltages across the secondary winding, near to the short circuit. In this regime, the flux within the magnetic circuit is very low, and the magnetizing current i_m is negligible (Fig. 3.3). With negligible i_m, the secondary current i_2 is very close to the $(N_1/N_2) \cdot i_1$, thus suppressing the measurement errors caused by the magnetizing current. Whatever the amplitude of the measured primary current, the rated power S_n of current transformers is 10–20 VA, with the external diameter close to 60 mm. The transformers are often toroidal, with only one primary turn. In most cases, the one primary turn is obtained by passing the line conductor through the inner opening of the torus.

The current transformers have a very large bandwidth. Even the frequency components of several tens of kilohertz get passed from the primary to the secondary circuit with acceptable amplitude errors and phase delays. Their problem resides at low frequencies. As the frequency of the measured current decreases, the flux within the magnetic circuit of the current transformer increases along with the magnetizing current. This increases the measurement error. Eventually, as the frequency reaches zero and the measured current becomes the dc current, the secondary current i_2 drops down to zero, and the operation of the current transformer is discontinued. The subsequent analysis of the low-frequency operation of current transformers is based on the equivalent circuit in Fig. 3.3, where the winding resistances are considered negligible.

In Fig. 3.3, the primary current and the primary impedances are referred to the secondary side. In steady-state operation and assuming that the primary current has the amplitude I_1 and the frequency ω, the phasor that represents the secondary current is obtained as

$$I_2(j\omega) = \frac{j\omega \cdot L_m}{j\omega \cdot L_m + j\omega \cdot L_{\gamma 2} + R_{SH}} \cdot \left(\frac{N_1}{N_2}\right) I_1(j\omega). \tag{3.2}$$

With toroidal shape of the transformer and with the uniform distribution of the secondary turns along the circumference, the flux caused by the secondary currents passes through the ferromagnetic torus, and the leakage is very low. Compared to L_m, the secondary leakage inductance $L_{\gamma 2}$ in (3.2) is negligible:

$$I_2(j\omega) \approx \frac{j\omega \cdot L_m}{j\omega \cdot L_m + R_{SH}} \cdot \left(\frac{N_1}{N_2}\right) I_1(j\omega). \tag{3.3}$$

The magnetizing current of toroidal transformers is lower than 2% of the rated current. In other words, the relative value of the magnetizing current is lower than 0.02. Therefore, the relative value of the impedance $\omega_F \cdot L_m$ obtained at the rated (line) frequency is larger than $1/0.02 = 50$. At the same time, the relative value of the voltage u_{SH} and the shunt resistance R_{SH} is lower than 1/10. At line frequency, the impedance $\omega_F \cdot L_m$ is some 500 times larger than the shunt resistor R_{SH}, and the phasor $I_2(j\omega)$ is very close to $I_1(j\omega) \cdot (N_1/N_2)$. As the frequency of the primary current reduces, the impedance $\omega_F \cdot L_m$ reduces and comes closed to R_{SH}. This increases the measurement error and introduces the difference between $I_2(j\omega)$ and $I_1(j\omega) \cdot (N_1/N_2)$. At $\omega = \omega_F/500$, the value of $\omega_F \cdot L_m$ reduces to R_{SH}, and the amplitude of the secondary current is reduced by 3 dB, while the phase error increases to $\pi/4$. In cases where the primary current is a dc current, the frequency is $\omega = 0$, as the secondary current drops down to zero.

The current to be measured in grid-side inverters is an ac current. Moreover, any dc component of the current that is injected into ac grid is harmful and should be avoided. Therefore, the question arises whether the current sensor aimed for the measurement of ac currents has to be capable to measure the dc currents. Even if the intention is that the grid-side inverters inject only ac currents into the grid, their imperfections such as uneven delays in PWM command lines and differences between the semiconductor power switches give a rise to a small, parasitic dc current. In order to remove the dc injection, it is necessary to use the current sensors capable of detecting the dc currents.

The inverters are also used in ac electrical drives, where they supply the ac currents into the stator windings of synchronous and asynchronous machines. In most cases, the drive has digital current controllers. The objective of current controllers is the injection of ac currents into the windings. Yet, in order to detect and suppress any parasitic dc currents, the current sensors used in electrical drives must be capable of measuring the dc currents as well as the ac. Therefore, it is not possible to use the conventional current transformers as the current sensors.

3.1.3   Rogowski Coils

Rogowski coil is a helical coil of insulated copper wire where the diameter of one turn is considerably smaller than the length. The coil does not extend in one direction only. Instead, it is flexible and capable of enveloping the current-carrying conductor (Fig. 3.4). Individual turns have the cross section S which is maintained along the coil length. The number of turns per unit length of the coil does not change either. Thus, the turn-density-per-length is constant. In Fig. 3.4, both terminals A and B are at the same end of the coil. For this to achieve, the wire end of the last turn is driven back through the center of the coil. Rogowski coil should be placed around the current-carrying conductor (i_1) in such way that the two ends of the coil eventually meet. The voltage u_{AB} across the coil ends is proportional to the first derivative of the measured current i_1.

Circular integral of the magnetic field H along the closed contour that envelops the current-carrying conductor is given in (3.4), where the contour C is any closed contour that encircles the current i_1, while dl is the infinitesimal length of the contour. In Fig. 3.4, the contour C is drawn as a circle with the current-carrying

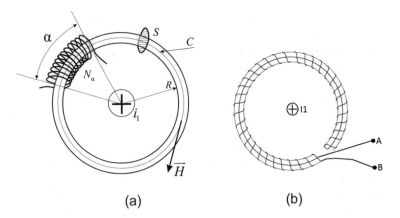

(a)   (b)

Fig. 3.4 Rogowski coil: (**a**) The cross section S of the coil is constant along the coil length. The number of turns per unit length is also constant. (**b**) Upon reaching the end of the coil, the wire returns to the beginning through the center of the coil. Thus, both terminals (A and B) are at the same end of the coil. In use, the coil envelops the current-carrying conductor (i_1)

conductor in the center, and the element dl is a vector collinear with the field H. In general case, the contour C does not have to be a circle which is centered around the conductor. The angle between vectors H and dl in (3.4) does not have to be zero:

$$\oint_C \vec{H} \cdot d\vec{l} = i_1. \tag{3.4}$$

Along the circular contour C in Fig. 3.4a, the field H is perpendicular to the surface S of each individual turn. Assuming that the diameter of the turn is considerably lower than the distance R from the current-carrying conductor, it is reasonable to consider the field strength H nearly constant along the surface S. Thus, the flux in each turn is $\Phi = \mu_0 \cdot H \cdot S$, where H is the magnetic field H in close vicinity of the considered turn. In cases where the field H is not perpendicular to the surface S, the flux is obtained by multiplying the factor $\mu_0 \cdot S$ and the scalar product of the vector H and the normal to the surface S.

The total flux of the coil is obtained by summing the fluxes of individual terms. In Fig. 3.4, the density of turns is

$$N' = \frac{N\alpha}{R \cdot \alpha}. \tag{3.5}$$

Along the coil length dl, there are d$N = N' \cdot$dl turns. The total flux ψ of the coil is obtained by integrating $\Phi \cdot$dN along the closed contour C:

$$\psi = \oint_C \Phi \cdot dN = \oint_C (\mu_0 \cdot S \cdot \vec{H} \cdot \vec{n}) \cdot N' \cdot dl. \tag{3.6}$$

Taking into account that

$$\vec{n} \cdot dl = d\vec{l},$$

the flux is obtained as

$$\psi = \mu_0 \cdot S \cdot N' \cdot \oint_C \vec{H} \cdot dl = \mu_0 \cdot S \cdot N' \cdot i_1. \tag{3.7}$$

The voltage u_{AB} across the coil is equal to

$$u_{AB} = R_C \cdot i_C + \frac{d\psi}{dt},$$

where R_C is resistance of the coil while i_C is the current within the coil. The voltage u_{AB} is brought to the detection circuit of Fig. 3.5. The resistance R_1 is larger than R_C by several orders of magnitude. Therefore, the voltage u_{AB} is very close to the first derivative of the coil flux (3.7) and, therefore, to the first derivative of the current i_1. For this reason, detection circuit (Fig. 3.5) is an integrator which provides the output u_{OUT}. In absence of R_2, the voltage u_{OUT} is obtained by integrating u_{AB}:

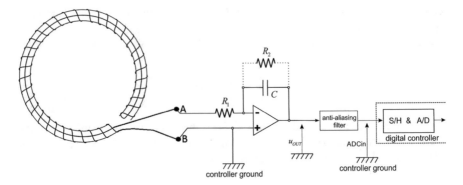

Fig. 3.5 Integration of the voltage detected by the Rogowski coil. Additional resistance R_2 is required in order to suppress the offset

$$u_{\text{OUT}} = -\frac{1}{C} \int \frac{u_{\text{AB}}}{R_1} \cdot dt = \frac{-\psi}{C \cdot R_1} = -\frac{\mu_0 \cdot S \cdot N'}{C \cdot R_1} \cdot i_1 = -k_x \cdot i_1. \tag{3.8}$$

In (3.8), the output voltage u_{OUT} of the detection circuit is proportional to the current i_1. In absence of any offset voltages and currents, and assuming that the initial values are $u_{\text{OUT}}(0) = 0$ and $i_1(0) = 0$, the output $u_{\text{OUT}}(t)$ is equal to $-k_x \cdot i_1(t)$ for both ac and dc currents.

Even a very small offset of the amplifier in Fig. 3.5 produces a buildup of u_{OUT} and obstructs the measurement. The dc gain of the $R_1 - C$ integrator is infinite. In other words, the gain $1/(R_1 \cdot C \cdot j\omega)$ becomes infinite as the frequency ω drops down to zero. Thus, it is necessary to make the integrator "leaky" by adding resistance R_2 in parallel with C. In this way, the dc gain of the leaky integrator becomes R_2/R_1, and the output response to the input offset V_{OIN} gets limited to $V_{\text{0OUT}} = R_2/R_1 \cdot V_{\text{OIN}}$.

The added resistance R_2 solves the problem of the integrator buildup, but it also brings in the errors in the current measurement. The error will be calculated by considering the measurement circuit in Fig. 3.5 with R_2 included and for the constant-frequency input i_1. Assuming that the measured current is $i_1 = I_m \cdot \cos(\omega t)$, it can be represented by the phasor $I_1(j\omega)$. The voltage $U_{\text{AB}}(j\omega)$ is obtained as

$$U_{\text{AB}}(j\omega) = j\omega\psi = j\omega \cdot \mu_0 \cdot S \cdot N' \cdot I_1(j\omega). \tag{3.9}$$

From the circuit in Fig. 3.5,

$$\begin{aligned} U_{\text{OUT}}(j\omega) &= \frac{-R_2}{R_1 \cdot (1 + j\omega \cdot C \cdot R_2)} U_{\text{AB}}(j\omega) \\ &= -\left(\frac{\mu_0 \cdot S \cdot N'}{R_1 \cdot C}\right) \cdot \frac{j\omega \cdot C \cdot R_2 \cdot I_1(j\omega)}{1 + j\omega \cdot C \cdot R_2} \\ &= -\left(\frac{j\omega \cdot C \cdot R_2}{1 + j\omega \cdot C \cdot R_2}\right) \cdot k_x \cdot I_1(j\omega). \end{aligned} \tag{3.10}$$

With large values of $j\omega \cdot \tau_2 = j\omega \cdot R_2 \cdot C$, the output voltage $U_{OUT}(j\omega)$ becomes very close to $-k_x \cdot I_1(j\omega)$. When the frequency ω of the measured current drops down and comes close to $1/\tau_2$, the amplitude error approaches -3 dB, while the phase error approaches $\pi/4$. Eventually, the measurement of dc current component is not possible. Therefore, Rogowski coil is used only in cases where it is necessary to measure just ac component of the current i_1, disregarding its dc component.

While it is rarely used for current-feedback purposes, Rogowski coil is a very practical solution for measuring the ac current in bulk conductors and rails that can hardly be encircled by conventional current transformers.

3.1.4 Hall Effect and Fluxgate Current Sensors

The measurement of the output current and the acquisition of the feedback signals are expected to transfer both ac and dc currents to the digital controller. Current transformers are robust, low-cost, and high bandwidth devices, but they lack the possibility to pass the dc currents and low-frequency currents. With dc currents in the primary winding, there are no electromotive forces and no currents in the secondary winding. The magnetomotive force of the primary winding does not get counterbalanced by the secondary currents, and the magnetic core of the current transformer becomes saturated. In all the operating conditions where the magnetizing current of the current transformer assumes considerable values, the basic relation $N_1 i_1 + N_2 i_2 = 0$ does not hold, and the measurement is erroneous.

In order to avoid core saturation, it is necessary to provide the means for preventing the magnetic saturation of the transformer core. This can be done by inserting a Hall element into conveniently designed air gap within the magnetic core, as illustrated in Fig. 3.6.

The Hall element is a semiconductor device that provides the signal proportional to the magnetic field. Shaded cuboid in Fig. 3.6 represents the Hall element inserted into the core slit. With appropriate polarization current, the Hall element provides the voltage u_H which is proportional to the induction B, $u_H = k_H \cdot B$. The core flux Φ_m is

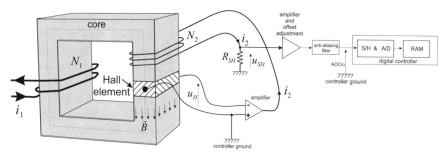

Fig. 3.6 Current-sensing transformer with Hall effect sensor placed in the air gap of the magnetic core

proportional to the induction B, which is perpendicular to the cross section S, $\Phi_m = S \cdot B$. At the same time, the flux Φ_m is obtained by dividing the magnetomotive force $F_m = N_1 i_1 + N_2 i_2$ by the equivalent magnetic resistance R_m of the magnetic circuit, $\Phi_m = F_m / R_m$. Assuming that the thickness of the Hall element is δ, and that the field strength H_{Fe} within the iron core is negligible, magnetic resistance is approximated by $R_m = \delta/(\mu_0 \cdot S)$. With $H_{Fe} \approx 0$, the core flux can be expressed by

$$\Phi_m = \frac{F_m}{R_m} = \frac{\mu_0 \cdot S}{\delta} (N_1 \cdot i_1 + N_2 \cdot i_2) = B \cdot S. \qquad (3.11)$$

From (3.11), the voltage u_H detected across the ends of the Hall element can be expressed in terms of the primary and secondary current:

$$u_H = k_H \cdot B = k_H \frac{\mu_0}{\delta} (N_1 \cdot i_1 + N_2 \cdot i_2) = k_{H1}(N_1 \cdot i_1 + N_2 \cdot i_2). \qquad (3.12)$$

In Fig. 3.6, the voltage u_H is brought to the amplifier which outputs the current i_2. The current creates the magnetomotive force $N_2 i_2$ which reduces the core induction B. With appropriate feedback actions and gains, the amplifier output i_2 keeps the magnetic core in zero-flux condition, namely, with $B = 0$ and $u_H = 0$. From (3.12), this leads to $N_1 i_1 + N_2 i_2 = 0$. Thus, the secondary current $i_2 = -(N_1/N_2) \cdot i_1$ remains proportional to the primary current for both ac and dc currents. The signal u_{SH} proceeds to the offset compensation, amplification, and anti-alias filtering before getting passed to the A/D converter.

At high frequency, the Hall effect current sensor of Fig. 3.6 operates on the same principles as the conventional current transformer. Thus, it is possible to measure the current at frequencies that exceed the bandwidth of the amplifier which outputs the secondary current. Namely, corrective action of the amplifier is indispensable only at very low frequencies. The Hall effect current sensors have a finite measurement. The manufacturing cost is increased by the need to perform a very precise slitting of the magnetic circuit.

The Hall element in Fig. 3.6 can be replaced by introducing additional, third winding which helps detecting whether the core flux is zero. In this case, there is no need for the air gap, and this simplifies the manufacturing of the core. The third winding is used to inject a small, high-frequency excitation at frequencies of several hundred kilohertz, considerably larger than the desired bandwidth of the current measurement chain. In response to the high-frequency excitation, the operating point in B-H diagram undergoes small high-frequency oscillations. High-frequency fluctuations of the inductance B and the magnetic field H can be detected by processing the voltage u_3 and the current i_3 of the third winding.

Due to nonlinear magnetizing characteristic of the ferromagnetic material, a small sinusoidal excitation i_3 produces distorted oscillations of the magnetic inductance B within the core. In cases where the core is polarized by the non-zero flux, the consequential non-sinusoidal components of B comprise significant low-order even harmonics. The phase of even harmonics depends on the flux polarity. The relevant even harmonics can be detected from the voltage u_3 across the third winding. Thus, the amplitude and sign of the core flux offset can be detected from the amplitude and phase of even harmonics detected from u_3.

In addition to fluxgate current sensors, the current measurement can be performed by magneto-optical current transformers (MOCT), devices that use fiber-optic cables and Faraday rotation. Current measurement by MOCT devices is convenient for the current measurement in high-voltage conductors within electrical power systems. Magnetic field H in close vicinity of the conductor is proportional to the electrical current. An optical cable has to be wrapped around the conductor. In this way, the optical cable is exposed to the field H. Through the Faraday effects, magnetic field H changes the polarization plane of the light that passes through the optical cable. Detection light detection and processing circuit with interferometer can be used to reconstruct the conductor current.

In grid-side power converters and electrical drives, most current sensors operate on the fluxgate principles, namely, as the current transformers with the auxiliary electronics and the compensating winding which keeps the transformer core in zero-flux conditions.

3.2 Analogue Filtering and Sampling

The signal obtained from the current sensor is usually the voltage signal where the range $u_{SH} \in [U_{min} \ldots U_{max}]$ corresponds to the change of the measured current $i_1 \in [-I_{max} \ldots +I_{max}]$. At the end of the feedback chain, the current feedback is represented by the finite wordlength number written in data memory of the digital controller. The analogue signal obtained from the sensor has to be passed through the analogue circuitry before being brought to the input of the sample-and-hold circuit of the A/D converter.

3.2.1 Gain and Offset Adjustment

With Hall effect current sensors resembling the one in Fig. 3.6, the associated electronic circuits are supplied from ± 15 Vdc, and the voltage u_{SH} across the shunt resistor often changes from $U_{min} = -6$ V up to $U_{max} = +6$ V. At the same time, most A/D converters are built as peripheral units of digital controllers, and their input voltage range is mostly $u_{ADC} \in [0 \ldots 3$ V]. In such cases, the output current levels of $-I_{max}$, 0, and $+I_{max}$ are represented by u_{ADC} voltage levels of 0, 1.5, and 3 V. Therefore, it is necessary to amplify (or attenuate) the signal and add the required offset voltage.

In cases where the voltage across the shunt resistor exceeds ± 3 V, the interface circuit can be passive, as shown in Fig. 3.7. The absence of active components, such as the operational amplifiers, removes their noise and offset. Resistances in Fig. 3.7 have to be selected in such way that the change of the shunt voltage u_{SH} from

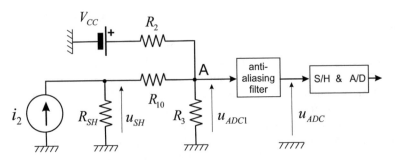

Fig. 3.7 Passive circuit for attenuation and offset correction of the feedback signals obtained from the current sensor

U_{min} to U_{max} produces the change of u_{ADC} from U_{Amin} to U_{Amax} (in most cases, from 0 to 3 V).

The shunt resistor R_{SH} and the current source i_2 can be replaced by the series connection of the ideal voltage source $u_{SH} = R_{SH} \cdot i_2$ and the series resistance R_{SH}. Connected in series, R_{SH} and R_{10} create an equivalent series resistance $R_1 = R_{SH} + R_{10}$. The voltage u_{ADC1} is fed to the anti-aliasing filter. Assuming that the input current of the filter is negligible, the sum of the currents that meet at the node A (i_{R1}, i_{R2}, i_{R3}) becomes zero:

$$\frac{u_{SH} - u_{ADC1}}{R_1} + \frac{V_{CC} - u_{ADC1}}{R_2} + \frac{-u_{ADC1}}{R_3} = 0. \tag{3.13}$$

For $u_{SH} = U_{max}$, the voltage u_{ADC1} should be equal to U_{Amax}, while for $u_{SH} = U_{min}$, u_{ADC1} should be equal to U_{Amin}. With $R_1 = R_3/a$ and $R_2 = R_3/b$,

$$\begin{aligned} a \cdot U_{max} + b \cdot V_{CC} - U_{Amax}(1 + a + b) = 0, \\ a \cdot U_{min} + b \cdot V_{CC} - U_{Amin}(1 + a + b) = 0. \end{aligned} \tag{3.14}$$

For $U_{min} = -6$ V, $U_{max} = +6$ V, $V_{CC} = +3$ V, and $u_{ADC1} \in [0 \dots 3$ V], the solution of (3.14) is $a = 1$ and $b = 2$. Thus, with $R_3 = 10$ kΩ, $R_1 = 10$ kΩ and $R_3 = 5$ kΩ. For $U_{min} = -3.3$ V and $U_{max} = +3.3$ V, the solution of (3.14) is $a = 10$ and $b = 11$. Notice in (3.14) that the proposed design flow defines the ratios R_3/R_1 and R_3/R_2. When selecting the resistance values, there is a degree of freedom. Namely, one can choose the value of R_3 at will, and the remaining values R_1 and R_2 will be determined by $R_1 = R_3/a$ and $R_2 = R_3/b$. Lower resistances reduce the effect of current disturbances on detected signals. On the other hand, the current drawn from the source V_{CC} is limited, and this limit sets the minimum value of R_2. The source V_{CC} is obtained from the ADC peripheral unit, it is decoupled from the digital noise, and it has a rather low current limit. In most cases, the three resistances of Fig. 3.7 are $10 \div 50$ times larger than the shunt resistance R_{SH}.

For the input voltage range below ±3 V, it is not possible to use the circuit of Fig. 3.7. Instead, the gain and offset adjustment has to be done by means of the circuit with an operational amplifier, such as the one given in Fig. 3.8.

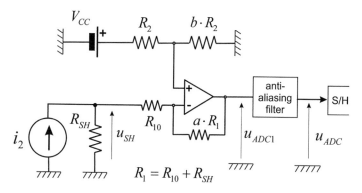

Fig. 3.8 Active interface circuit for the gain and offset correction of the feedback signals obtained from the current sensor

In the interface circuit of Fig. 3.8, the input current of the operational amplifier and the input current of the anti-aliasing filter are negligible. The input range $u_{SH} = R_{SH} \cdot i_2 \in [U_{min} \ldots U_{max}]$ should be mapped to the output range $u_{ADC1} \in [U_{Amin} \ldots U_{Amax}]$. In most cases, $u_{ADC1} \in [0 \ldots 3 \text{ V}]$. Notice in Fig. 3.8 that the proposed amplifier inverts the signal. Thus, the input U_{max} will bring the voltage u_{ADC1} to the lower end, to U_{Amin}, while the input U_{min} results in $u_{ADC1} = U_{Amax}$. The values of a and b can be obtained from

$$
\begin{aligned}
(1+b) \cdot (a \cdot U_{max} + U_{Amin}) &= (1+a) \cdot b \cdot V_{CC}, \\
(1+b) \cdot (a \cdot U_{min} + U_{Amax}) &= (1+a) \cdot b \cdot V_{CC}.
\end{aligned}
\tag{3.15}
$$

For $U_{min} = -6$ V, $U_{max} = +6$ V, $V_{CC} = +3$ V, and $u_{ADC1} \in [0 \ldots 3 \text{ V}]$, the solution of (3.15) is $a = 1/4$ and $b = 2/3$. In Fig. 3.8, the choice of R_1 and R_2 is free, while the remaining two resistances are obtained as $a \cdot R_1$ and $b \cdot R_2$. Resistances R_1 and R_2 should be $10 \div 50$ times larger than R_{SH}.

The signal u_{ADC1} in Figs. 3.7 and 3.8 is scaled to fit the input voltage range of the A/D converter. Before the conversion, it is necessary to apply the anti-aliasing filter, the design of which is the subject of the subsequent discussion.

Question (3.1) The output current has an arbitrary waveform and changes between $\pm I_{max} = \pm 500$ A. It is measured by the Hall effect current sensor of Fig. 3.6. The transformation ratio $N_2:N_1$ can be either 1000:1 or 5000:1. Desirable range of the shunt voltage is $u_{SH} = \pm 6$ V. The power dissipation on the shunt resistor should not exceed 1 W. It is necessary to select the transformation ratio and the value of the shunt resistance.

Answer (3.1) With the transformation ratio 1000:1, the secondary current i_2 may reach 500 mA. If, at the same time, the shunt voltage reaches 6 V, the power dissipation of the shunt reaches 3 W and exceeds the limit of 1 W. Thus, the transformation ratio has to be 5000:1. Corresponding secondary current is

$i_2 = 500 \text{ A}/5000 = 100 \text{ mA}$. The shunt resistance is $R_{SH} = 6 \text{ V}/100 \text{ mA} = 60 \text{ }\Omega$. The power dissipation is $100 \text{ mA} \cdot 6\text{V} = 0.6 \text{ W} < 1 \text{ W}$.

Question (3.2) The output current is measured by the Hall effect current sensor which provides the shunt voltage $u_{SH} = \pm 5 \text{ V}$. The shunt resistance is 50 Ω. The input voltage range of the A/D converter is $u_{ADC1} \in [0 \dots 3 \text{ V}]$. It is necessary to select the resistances R_{10}, R_2, and R_3 of the interface circuit in Fig. 3.7. Resistances should be as low as possible. Maximum permissible current of the source $V_{CC} = 3 \text{ V}$ is 5 mA.

Answer (3.2) With $R_1 = R_3/a$ and $R_2 = R_3/b$, design equations are reduced to the expression (3.14). For the given parameters, the system of two equations is solved for $a = 3/2$ and $b = 5/2$. The largest current of the source V_{CC} is obtained with $u_{ADC1} = 0$, and it is equal to V_{CC}/R_3. Therefore, the lowest value of R_2 is 3 V/ 5 mA = 600 Ω. This results in $R_3 = R_2 * b = 1500 \text{ }\Omega$ and $R_1 = 1000 \text{ }\Omega$. Resistance R_{10} is equal to $R_1 - R_{SH} = 950 \text{ }\Omega$.

Question (3.3) The output current is measured by the Hall effect current sensor which provides the shunt voltage u_{SH} which changes from -4 V to $+4$ V. The input voltage range of the A/D converter is $u_{ADC1} \in [0 \dots 3 \text{ V}]$. The shunt resistance is equal to $R_{SH} = 100 \text{ }\Omega$. It is necessary to select the resistances R_{10}, R_2, and R_3 of the interface circuit in Fig. 3.7. Resistance R_1 should be 20 times larger than R_{SH}.

Answer (3.3) The values a and b are obtained from (3.14), $a = 3$, $b = 4$. From $R_1 = 20 \cdot 100 \text{ }\Omega = 2 \text{ k}\Omega$, $R_{10} = 1950 \text{ }\Omega$. With $R_1 = R_3/a$, $R_3 = a \cdot R_1 = 6 \text{ k}\Omega$. From $R_2 = R_3/b$, $R_2 = 1500 \text{ }\Omega$.

Question (3.4) The output current is measured by the Hall effect current sensor which provides the voltage u_{SH} which changes from -3 to $+3$ V. The internal resistance of the voltage source u_{SH} is negligible. The input voltage range of the A/D converter is $u_{ADC1} \in [0 \dots 3 \text{ V}]$. It is necessary to select the resistances of the interface circuit in Fig. 3.8, where $V_{CC} = +3 \text{ V}$. None of the resistances should be lower than 1000 Ω.

Answer (3.4) The values a and b are obtained from (3.15), $a = 1/2$, $b = 1/2$. In order to keep $a \cdot R_1$ and $b \cdot R_2$ above 1 kΩ, it is necessary to choose $R_1 = R_2 = 2 \text{ k}\Omega$.

Question (3.5) The output current is measured by the Hall effect current sensor that has internal amplifier which provides the voltage signal u_{SH} that changes from 1.5 to 4 V. The internal resistance of the voltage source u_{SH} is negligible. The input voltage range of the A/D converter is $u_{ADC1} \in [0 \dots 3 \text{ V}]$. It is necessary to select the resistances of the interface circuit in Fig. 3.8, where $V_{CC} = +3 \text{ V}$. None of the resistances should be lower than 1000 Ω.

Answer (3.5) The values a and b are obtained from (3.15), $a = 1.2$, $b = 2.66$. With $R_1 = R_2 = 1 \text{ k}\Omega$, the remaining two resistances are obtained as $a \cdot R_1$ and $b \cdot R_2$.

3.2.2 Analogue-to-Digital Conversion

The signal u_{ADC} of Fig. 3.8 is brought to the input of the A/D converter. The A/D converter is one of the peripheral units of the digital signal controller which receives the measured analogue signals and converts them into digital form, that is, it turns them into numbers. Further on, such numbers are used as the feedback in software-implemented current controllers. Amplitude quantization of A/D converter is illustrated in Fig. 3.9.

The A/D converter provides an N-bit equivalent of the input analogue signals, where N is most frequently $8 < N < 16$. The N-bit equivalent is often called the ADC code. Relation between the analogue signal and the ADC code is given in Fig. 3.9a. For the purpose of argument, the figure provides an example obtained with $N = 4$. The smallest detectable change of the input voltage is $\Delta u_Q = U_{\text{MAX}}/2^N$. As the input voltage u_{ADC} sweeps from 0 to U_{MAX}, the ADC code changes from 0000 to 1111. The process of converting the analogue signal into number implies the lack of information. In cases where the ADC code is equal to M, the input signal u_{ADC} resides between $M \cdot \Delta u_Q$ and $(M + 1) \cdot \Delta u_Q$. Thus, the ADC code does not provide the information on the exact value of u_{ADC} between the two thresholds.

The errors introduced by the amplitude quantization are illustrated in Fig. 3.9b. The figure presents the time change of the input signal u_{ADC} and the corresponding ADC code. The ADC code is represented by $M \cdot \Delta u_Q$. The difference between the two traces is the quantization error. In Fig. 3.9b, the code M is determined as the integer part of $u_{\text{ADC}}/\Delta u_Q$, and the error changes between 0 and Δu_Q. By introducing an offset of $\Delta u_Q/2$, the error is brought to the strip $\pm \Delta u_Q/2 = \pm U_{\text{MAX}}/2^{N+1}$. The number of bits N in the ADC code is often called the A/D resolution. Thus, a 12-bit resolution A/D converter has $\Delta u_Q = U_{\text{MAX}}/4096$.

In order to reduce the amplitude-quantization error, it is of interest to increase the resolution N. Yet, there are two reasons to keep the A/D resolution low:

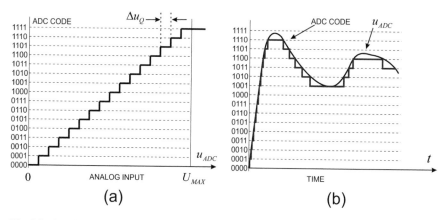

Fig. 3.9 Amplitude quantization in A/D converter

- With larger values of N, it takes more time to complete the analogue-to-digital conversion. This introduces delays into the feedback loop. In cases where a single A/D converter processes several analogue inputs, selected by means of an analogue multiplexer, this delay can be significant.
- Along with the input signal u_{ADC}, the A/D converter picks up the conducted and irradiated noise. The noise is generated by the switching transients across the semiconductor power switches and also by the switch-mode power supplies. At the same time, an A/D peripheral unit is often embedded into the same chip with a microprocessor or a digital signal processor. Even a single-chip A/D converter is surrounded by digital circuits where the TTL-level signals commutate at the clock frequency. This intrinsic *digital noise* can be reduced by decoupling the A/D converter supply and control logic from the fast-switching digital devices. Yet, notwithstanding the art of decoupling and the skills in designing low-noise layout, it is rather difficult to suppress the noise within u_{ADC} signals. Fending off the non-conventional solutions, it is difficult to suppress the noise below 1 mV. Considering the input voltage range of $U_{MAX} = 3$ V, the noise is comparable with one quantization step of the 12-bit converter, where $\Delta u_Q = U_{MAX}/2^{12} \approx 0.73$ mV.

Therefore, most embedded A/D converters are $N = 12$-bit devices, capable of converting an analogue value into number in less than 200 ns.

In cases requiring precision better than $\Delta u_Q = 0.73$ mV, it is a common practice to put to use the A/D throughput and acquire a number of consecutive samples of the same signal. In certain conditions, an average value obtained from such train of samples would represent the analogue input u_{ADC} with resolution better than $U_{MAX}/2^{12}$. This technique is called the *oversampling*.

3.2.3   Sampling Process

Digital current controller compares the current feedback to the reference current, derives the current error, and then calculates the voltage command capable of reducing the error and eventually making the current track the reference with no errors.

The process of feedback acquisition and the execution of the software-implemented controller introduce finite time delays Δt_{FB} and Δt_{EXE}. Therefore, it takes $\Delta t = \Delta t_{FB} + \Delta t_{EXE}$ to complete the feedback acquisition processes and calculate the new voltage command. After that, the hardware resources are available to initiate the next acquisition/calculation sequence. Thus, the digital implementation of the controller has an inherent discrete-time nature, namely, control sequences keep repeating with the period T, also called the sampling period.

The voltage actuator of the current controller is the PWM inverter. Within each switching period $T_{PWM} = 1/f_{PWM}$, the inverter outputs the voltage pulse. The pulse width t_{ON} of the voltage pulse determines the average value of the output voltage, the

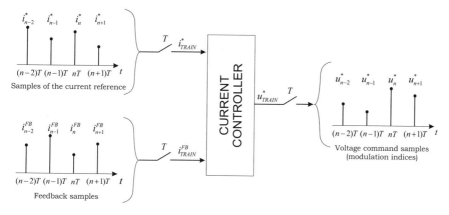

Fig. 3.10 Time-discrete nature of digital controllers

actual driving force of the control loop. The modulation index $m = t_{ON}/T_{PWM}$ assumes the role of the voltage command. According to the analysis given in Sect. 2.1.5, the position of both rising and falling edges of each voltage pulse can be affected by different modulation indices, providing the means to control the average voltage in each half-period of the PWM, each $T = T_{PWM}/2$. Therefore, the digital current controller has the possibility to apply one voltage command in each sampling period T. For that reason, the processes of feedback acquisition and calculation of the new voltage reference are repeated each T.

Discrete-time nature of the current controller is illustrated in Fig. 3.10. At each instant $t = n \cdot T$, the current error is calculated as the difference between the newly acquired feedback sample i_n^{FB} and the reference i_n^*. Based on the current error, the current controller calculates the voltage reference u_n^*. Thus, the digital current controller of Fig. 3.10 receives the train of equidistant feedback samples i_n^{FB} and the train of reference values i_n^*, and it generates the output train of the voltage command samples u_n^*.

The current feedback is provided in the form of an analogue signal u_{ADC}, which is brought to the input of the A/D converter (Fig. 3.8), namely, to the sample-and-hold (S/H) circuit. Capacitor within the S/H circuit gets charged to the voltage u_{ADC} before the process of A/D conversion begins. For the proper, error-free operation of the A/D converter, it is required that the analogue signal being converted does not change in the course of the conversion. For this reason, the S/H capacitor is disconnected from the input signal, while the conversion is in progress. Thus, the voltage brought to the A/D converter does not change during the conversion. The signal $u_{ADC}(t)$ is sampled at instants nT, and the values $u_{ADC}(nT)$ are retained at the S/H capacitor and converted into N-digit binary numbers. Corresponding sampling circuit and the voltage across the sampling capacitor are shown in Fig. 3.11.

The sampling circuit of Fig. 3.11 is usually embedded within a single chip with the A/D converter, other peripherals, and CPU of the digital controller. The switch SW is the analogue switch which closes at the sampling instants. It has to remain

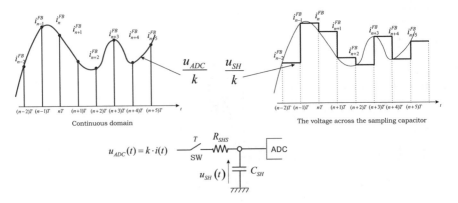

Fig. 3.11 Sampling circuit and the voltage across the sampling capacitor

closed for several time constants $\tau_{SH} = R_{SH} \cdot C_{SH}$, so that the voltage u_{SH} can exponentially approach the input voltage u_{ADC} with an error inferior to $\Delta u_Q = U_{MAX}/2^N$. With $N = 12$-bit A/D converter, the switch SW has to remain closed for $8 \div 9\,\tau_{SH}$. The value Δu_Q corresponds to one least significant bit (LSB) of the ADC code. Thus, the errors inferior to 1 LSB are considered acceptable. The value of R_{SH} in Fig. 3.11 includes the equivalent resistance of the external circuits (Fig. 3.8) that supply to voltage u_{ADC} to the sampling circuit. For that reason, the interval of time when the analogue switch SW is closed is usually programmable, and it can be adjusted by setting the appropriate code in configuration registers of the ADC peripheral unit. In most cases, it is possible to charge the sampling capacitor in less than 100 ns. Once the capacitor C_{SH} is charged and the switch SW opened, the voltage u_{SH} does not change until the next sampling instant.

Typical PWM periods are well above $T_{PWM} = 50\,\mu s$, with sampling periods above $T = T_{PWM}/2 = 25\,\mu s$. The processes of charging the sampling capacitor and performing the A/D conversion complete in less than 300 ns. Therefore, it is reasonable to assume that the A/D converter captures the feedback samples i^{FB} (nT) right at the sampling instants $t = nT$, and it converts them into corresponding numbers. The train of pulses is shown in Fig. 3.12.

Each current sample can be modeled by $i_n^{FB} \cdot \delta(t - nT))$, by the product obtained by multiplying Dirac delta function time-shifted by nT. The model is illustrated in Fig. 3.13. When the surface of the shaded area is equal to the value of the sample, and ε reduces to zero, the shaded area assumes the form of the Dirac delta function.

With mathematical representation of Fig. 3.13, the train of samples can be expressed as $i_D^{FB}(t)$ in (3.16). The samples are brought into the digital controller (Fig. 3.10) and used to calculate the output, the sample train of the voltage references suited to suppress the current error:

$$i_D^{FB}(t) = \sum_{k=0}^{+\infty} i^{FB}(kT) \cdot \delta(t - kT). \tag{3.16}$$

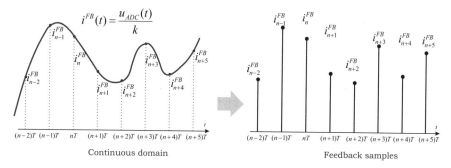

Continuous domain   Feedback samples

Fig. 3.12 Conversion of the analogue feedback signal into the train of samples

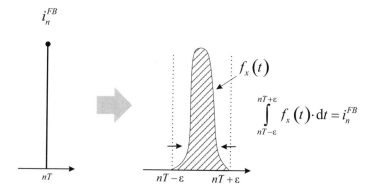

Fig. 3.13 Mathematical representation of the sample acquired at instant nT

In addition to the information loss due to amplitude quantization in A/D converter, the process of sampling continuous-time signal $i^{FB}(t)$ and obtaining the pulse train of (3.16) introduces an additional lack of information due to time-quantization effects. The pulse train $i_D^{FB}(t)$ of (3.16) defines the values at instants $t = kT$, where T is the sampling period, but it cannot be used to reconstruct the continuous-time function $i^{FB}(t)$ in general case. Namely, the samples continuous-time signal $i^{FB}(kT)$ and $i^{FB}((k + 1)T)$ do not provide sufficient information to reconstruct $i^{FB}(t)$ within the interval $[kT \ldots (k + 1)T]$, unless the continuous-time function does not meet specific criteria.

In closed-loop systems, the objective of the control is to suppress the errors in the region of low frequencies, from dc up to the bandwidth frequency f_{BW}, usually defined as the frequency of the input sinusoidal excitation where the attenuation of the consequential output is 3 dB (1/sqrt(2)). The errors that pulsate at frequencies cannot be suppressed due to performance limits of the controller and the actuator. Therefore, the high-frequency errors caused by the time quantization are of no interest if their frequency resides well beyond the bandwidth frequency f_{BW}. On the other hand, the sampling errors at low frequencies could introduce significant

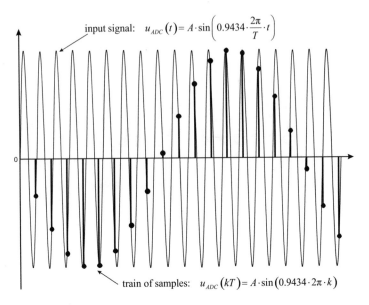

Fig. 3.14 A low-frequency alias caused by erroneous sampling

disturbance and deteriorate the operation of the controller. An example of such errors is illustrated in Fig. 3.14. The input signal is a sinusoid with the period some 6% larger than the sampling period. Thus, the sampling instants are sliding gradually along the slopes of the sinusoidal waveform. As a consequence, the resulting train of sample depicts the slow-changing sinusoidal form at the frequency f_a which is some 16 times lower than the frequency of the original signal $u_{ADC}(t)$.

When the pulse train of Fig. 3.14 enters the current, controller receives the error signal of the frequency f_a, which does not exist in the original input signal. This frequency component is also called an *alias*. In order to suppress the nonexistent error at the frequency f_a, the controller will output the control output u^* that pulsates at the alias frequency. As a consequence, the controlled output current would obtain the actual error at the alias frequency. For that reason, it is necessary to organize the sampling process in such way that the alias-frequency components do not appear. For that to achieve, it is necessary to design and implement an analogue anti-aliasing filter before bringing the input signal to the A/D converter.

3.2.4   The Alias-Free Sampling

In order to analyze the alias frequencies and design the anti-alias filter, it is necessary to study the frequency content of the sample train (3.16). This can be done by deriving the Laplace transformation $i_D^{FB}(s)$ of the original $i_D^{FB}(t)$ and introducing

$s = j\omega$ in order to find the amplitude of individual frequency components. By introducing the function $\Lambda(t)$ as the sum of the delta functions, the train of samples can be expressed as the product of the original $i^{FB}(t)$ and the function $\Lambda(t)$ (3.17). The Laplace transformation of the sample train $i_D^{FB}(t)$ is given in (3.18):

$$i_D^{FB}(t) = i^{FB}(t) \cdot \sum_{k=0}^{+\infty} \delta(t - kT) = i^{FB}(t) \cdot \Lambda(t),$$

$$\Lambda(t) = \sum_{k=0}^{+\infty} \delta(t - kT). \tag{3.17}$$

$$L\left(i_D^{FB}(t)\right) = \int_{0^-}^{+\infty} i^{FB}(t)\Lambda(t)e^{-st}dt = \sum_{k=0}^{+\infty} i^{FB}(kT) \cdot L(\delta(t - kT)). \tag{3.18}$$

The Laplace transformation of the time-shifted delta function $\delta(t - kT)$ is equal to e^{-skT}. Therefore, Eq. (3.18) can be rewritten as

$$L\left(i_D^{FB}(t)\right) = I_D^{FB}(s) = \sum_{k=0}^{+\infty} i^{FB}(kT) \cdot e^{-ksT}. \tag{3.19}$$

The above expression provides the Laplace transformation $I_D^{FB}(s)$ of the sample train $i_D^{FB}(t)$, but it does not relate $I_D^{FB}(s)$ to the Laplace transformation $I^{FB}(s) = L(i^{FB}(t))$ of the original feedback signal $i^{FB}(t)$. The study of relation between the complex images $I_D^{FB}(s)$ and $I^{FB}(s)$ will provide the conclusions required to design the anti-aliasing filter. Only the conclusions of the subsequent analysis are required for the filter design. Therefore, only the main steps of $I_D^{FB}(s)$ calculation are outlined, leaving the rigorous proofs to attentive readers with developed skills in complex analysis. Others could proceed by considering the expression (3.25).

The sample train $i_D^{FB}(t)$ is expressed as the product of two time-domain functions, $i_D^{FB}(t) = i^{FB}(t) \cdot \Lambda(t)$. The Laplace transform of the product of two time-domain function is equal to the correlation of their complex images in s-domain. If $I_D^{FB}(s)$ is the Laplace transformation of the sample train, $I^{FB}(s)$ is the Laplace transformation of $i^{FB}(t)$, while $\Pi(s)$ is the Laplace transformation of $\Lambda(t)$, then the complex image $I_D^{FB}(s)$ of the sample train is calculated from

$$I_D^{FB}(s) = \frac{1}{(2\pi j)} \int_{\gamma-j\infty}^{\gamma+j\infty} \Pi(s - p) \cdot I^{FB}(p) \cdot dp. \tag{3.20}$$

The Laplace transformation of $\Lambda(t)$ is given in (3.21), and the convolution integral assumes the form (3.22) and provides the complex image $I_D^{FB}(s)$. The constant γ in Eqs. (3.20) and (3.22) defines the vertical line in s-plane which separates all the poles of $\Pi(s)$ from the imaginary axis. Specific details related to the Laplace transformation and convolution integrals in s-domain can be found in textbooks on complex analysis.

$$\Pi(s) = \int_{0^-}^{+\infty} \sum_{k=0}^{+\infty} \delta(t - kT) \cdot e^{-st} dt = \sum_{k=0}^{+\infty} e^{-ksT} \tag{3.21}$$

$$|e^{-sT}| < 1 \Rightarrow \Pi(s) = \frac{1}{1 - e^{-sT}}$$

$$I_D^{FB}(s) = \frac{1}{(2\pi j)} \int_{\gamma-j\infty}^{\gamma+j\infty} \frac{I^{FB}(p)}{1 - e^{-(s-p)T}} dp. \tag{3.22}$$

The line γ can be extended to a closed curve by adding a half-circle that has an infinite radius and resides in the right half-plane. In this way, one obtains a negatively oriented simple closed curve C. The complex image $I^{FB}(s)$ of practical input signals $i^{FB}(t)$ does not have any poles in the right half-plane. On the other hand, all the poles of $\Pi(s - p)$ are encircled by the curve C. These poles are $p_n = s + j \cdot n \cdot 2\pi/T = s + j \cdot n \cdot \Omega$, where $n = 0, \pm 1, \pm 2, \pm 3, \ldots$, while $\Omega = 2\pi/T$.

According to Cauchy's residue theorem, the line integral of the function $\Pi(s - p) \cdot I^{FB}(p)$ around C is equal to the sum of residues of $\Pi(s - p) \cdot I^{FB}(p)$ at the poles $p_n = s + j \cdot n \cdot \Omega$, multiplied by $-2\pi j$. The sign $(-)$ is due to negative orientation of the curve C. The residue at the pole p_n is obtained in (3.23), while the sum of residues is given in (3.24).

$$\lim_{p \to s+j \cdot n \cdot \Omega} \left[(p - s - j \cdot n \cdot \Omega) \cdot \frac{I^{FB}(p)}{1 - e^{-(s-p)T}} \right]$$

$$= I^{FB}(s + j \cdot n \cdot \Omega) \cdot \lim_{p \to s+j \cdot n \cdot \Omega} \left[\frac{(p - s - j \cdot n \cdot \Omega)}{1 - e^{-(s-p)T}} \right]$$

$$= I^{FB}(s + j \cdot n \cdot \Omega) \cdot \lim_{p \to s+j \cdot n \cdot \Omega} \left[\frac{1}{\frac{d}{dp}\left(1 - e^{-(s-p)T}\right)} \right] \tag{3.23}$$

$$= \frac{1}{T} I^{FB}(s + j \cdot n \cdot \Omega).$$

$$\frac{1}{(2\pi j)} \oint_C \frac{I^{FB}(p)}{1 - e^{-(s-p)T}} dp = \frac{1}{T} \sum_{n=-\infty}^{n=+\infty} I^{FB}(s + j \cdot n \cdot \Omega). \tag{3.24}$$

Desired complex image $I_D^{FB}(s)$ of the sample train is obtained from (3.22), as the convolution integral along the line γ. The line integral in (3.24) comprises the result (3.22), but it also includes the line integral along the half-circle that extends the line γ into the closed curve C. The half-circle has an infinite radius, and it is located in the right half-plane. The line integral I_{HC} along the half-circle can be calculated by replacing $p = R \cdot e^{j\theta}$ into the function $\Pi(s - p) \cdot I^{FB}(p)$, where θ goes from $\pi/2$, passes 0, and reaches $-\pi/2$, while $R \to +\infty$ is the half-circle radius.

With all the practical input signals $i^{FB}(t)$, the complex image $I^{FB}(s)$ can be expressed as the quotient of two real-coefficient polynomials where the order of the numerator is lower than the order of the denominator. Under circumstances, and according to the initial-value theorem of the Laplace transformation, the initial value of the time-domain function $i^{FB}(0^+)$ can be found as the limit value of $p \cdot I^{FB}(p)$ when $p \to +\infty$. Therefore, The line integral I_{HC} along the half-circle becomes $i^{FB}(0^+)/2$. Thus, the complex image of the sample train $i_D^{FB}(t)$ is given in (3.25). The spectral content of the train of samples can be obtained by replacing s by $j\omega$.

$$I_D^{FB}(s) = \frac{1}{T} \sum_{n=-\infty}^{n=+\infty} I^{FB}(s + j \cdot n \cdot \Omega) + \frac{1}{2} i^{FB}(0^+). \qquad (3.25)$$

The initial value $i^{FB}(0^+)$ of the feedback signal can be neglected without the lack of generality. With $n = 0$, the element $I^{FB}(j\omega)$ of the sum is the spectrum of the time-domain signal $i^{FB}(t)$, that is, the spectrum feedback signal in so-called analogue domain. The remaining elements of the sum in (3.25) represent the images of the same spectrum $I^{FB}(j\omega)$, shifted by $j \cdot n \cdot \Omega$ along the frequency axis, where $n = 0, \pm 1, \pm 2, \pm 3, \ldots$, while $\Omega = 2\pi/T$.

The result (3.25) is used to obtain the spectrum of the train of samples from the spectrum $I^{FB}(j\omega)$ of the time-domain (analogue) signal $i^{FB}(t)$. The upper trace in the figure represents the analogue feedback signal $i^{FB}(t)$. The sample spectrum $I^{FB}(j\omega)$ is drawn as a triangular shape which drops to zero at frequency $\omega < \Omega/2$, where $\Omega = 2\pi/T$ is the sampling frequency. Thus, $I^{FB}(j\omega)$ is equal to zero for any frequency that exceeds $\Omega/2$, one half of the sampling frequency. The trace in the middle represents the spectrum of the train of samples $i_D^{FB}(t)$. It is obtained by repeating the scaled spectrum $I^{FB}(j\omega)$ along the frequency axis. The repetition period is Ω.

The lower trace in Fig. 3.15 corresponds to the spectrum obtained by applying a low-pass filter to reconstruct the time-domain signal from the sampled data. Assuming that the low-pass filter does not introduce any amplitude or phase errors within the frequency range of interest, then the spectrum of the lower trace is equal to the scaled but otherwise unaltered original spectrum $I^{FB}(j\omega)$. Thus, in the case considered in Fig. 3.15, the sampling process and time quantization do not cause any loss of information. In other words, it is possible to reconstruct the original signal $i_{FB}(t)$ from the train of samples with no errors.

The spectra shown in Fig. 3.16 are drawn in the same manner as in Fig. 3.15. This time, the relevant traces are obtained with $I^{FB}(j\omega)$ which extends beyond $\Omega/2$. In the middle trace of the figure, there is considerable overlap between the neighboring images. An attempt to reconstruct the original signal by applying the low-pass filter results in the lower trace, where the side frequency bands of $I^{FB}(j\omega + j\Omega)/T$ and $I^{FB}(j\omega - j\Omega)/T$ overlap with the spectral image $I^{FB}(j\omega)/T$, positioned at the origin. Under circumstances, it is not possible to reconstruct the original signal. In other words, the sampling process introduces the errors. In order to make the errors more obvious, it is of interest to consider the frequency component of $I^{FB}(j\omega)$ at frequency $\Omega - \Delta\Omega$, denoted by (A) in Fig. 3.16. Due to the overlapping, the frequency

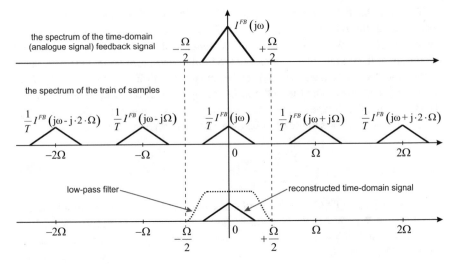

Fig. 3.15 The spectrum of the analogue signal $i^{FB}(t)$ (the upper trace) and the spectrum of the train of samples $i_D^{FB}(t)$ (in the middle). The lower trace corresponds to the spectrum obtained by using a low-pass filter to reconstruct the time-domain signal from the sampled data

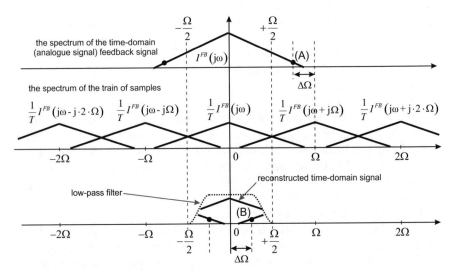

Fig. 3.16 The spectrum $I^{FB}(j\omega)$ of the analogue signal $i^{FB}(t)$ (the upper trace), the spectrum I_D^{FB} $(j\omega)$ of the train of samples $i_D^{FB}(t)$ (the middle), and the reconstructed spectrum (the lower trace) obtained with $I^{FB}(j\omega)$ which extends beyond $\Omega/2$

component $\Omega - \Delta\Omega$ appears within the reconstructed spectrum in the lower trace, as the frequency component $\Delta\Omega$, denoted by (B). The difference $\Delta\Omega$ can be just a small fraction of the sampling frequency. In such case, a high-frequency component of i^{FB} (t) could provoke a low-frequency component in the reconstructed signal, the component which does not exist in the original signal. For this reason, the train of samples in Fig. 3.14 comprises a false low-frequency signal. Such false signals are also called the alias signals. For the proper operation of the controller, it is essential to prevent the appearance of the alias components.

The mechanism where the high-frequency component of the original signal causes a low-frequency alias within the train of samples and the reconstructed signal is illustrated in Fig. 3.17. Assuming that the sampling frequency is $1/T = 10$ kHz, and that the original signal $i^{FB}(t)$ comprises a single frequency component at $\Omega - \Delta\Omega = 9990$ Hz, the spectrum $I^{FB}(j\omega)$ corresponds to the one shown as (A) in Fig. 3.17. The consequential spectrum $I_D^{FB}(t)$ of the sample train is shown as (B).

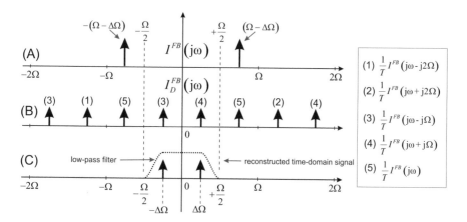

Fig. 3.17 Illustration of the case where the high-frequency component of the original signal causes a low-frequency alias within the reconstructed signal

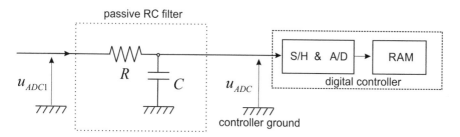

Fig. 3.18 Passive anti-aliasing R-C filter

The reconstructed spectrum (C) has the frequency component at $\Delta\Omega = 10\,\text{Hz}$, which does not exist in the original signal.

According to the results obtained above, the alias frequency component in reconstructed signal appears in cases where the original signal $i^{FB}(t)$ comprises the frequency components above one half of the sampling frequency ($\Omega/2$). Thus, in order to avoid the sampling errors and alias components, the analogue signal brought to the sampling circuit must not have any frequency component beyond $\Omega/2$ limit. With $I^{FB}(j\omega) = 0$ for any $|\omega| > \Omega/2$, the sampling process does not introduce any errors, it does not cause the lack of information, and it leaves the possibility to reconstruct the original signal from the train of samples. A more rigorous and formal statement of the above findings has been introduced and proved by Vladimir A. Kotelnikov. Theoretical work on the sampling process is largely associated with Claude Shannon (thus, Shannon's sampling theorem) and, to some extent, to Harry Nyquist. In contemporary literature, the frequency $\Omega/2$ is also called the Nyquist frequency.

In short terms, an error-free sampling requires that the analogue input signal $i^{FB}(t)$ does not have any frequency component above one half of the sampling frequency ($\Omega/2$). For this to achieve, it is necessary to use the low-pass filters prior to bringing the signal to the sampling circuit. This filter operates on time-domain signals, and it operates in analogue domain. Since its purpose is the suppression of alias frequencies, it is also called the anti-alias filter.

The feedback acquisition chain in digital control systems comprises sensors, analogue amplifiers, anti-aliasing filters, A/D converters and digital filters that process the sample train provided by the A/D converter. Design process starts from the desired bandwidth f_{BWA} of the feedback acquisition chain. All the relevant dynamic phenomena of the reside at frequencies below f_{BWA}, while most of the noise and high frequency dynamics remain at frequencies, well above f_{BWA}. The noise and high frequency dynamic phenomena cannot be controlled nor suppressed due to the bandwidth limits of controllers and actuators. Said phenomena are also called the unmodeled dynamics, and they are left to decay with their own damping and pace. The unmodeled dynamics should be suppressed and prevented from entering the system through the feedback chain, as the high frequency noise introduces sampling errors and impairs the operation of the closed loop systems.

The first and the key design step is selection of the sampling frequency. The former analysis proves that any frequency components above one half of the sampling frequency ($1/2/T$) have to be removed from the input signals. This condition is met by applying the analogue pre-filter called the anti-aliasing filter, discussed in the subsequent subsections. For practical reasons, such filter will also introduce some amplitude and phase distortions in the frequency range below $1/(2T)$. In order to reduce the amplitude and phase errors, the choice of the sampling frequency ($1/T$) has to keep the most important dynamic features of the system far below $1/(2T)$, in the frequency range where the phase and amplitude errors introduced by anti-aliasing filters are negligible. Design choice $T = 1/(20f_{BW})$ puts the sampling frequency an order of magnitude above $1/(2T)$, reducing in this way the impact of anti-aliasing filters. The final design values depend on the characteristics of the A/D converters, dynamic properties of the controlled object, the wordlength, the noise content, and other practical details.

Once the sampling frequency is determined, design proceeds with setting the anti-aliasing filter, discussed in the subsequent subsections. The sampling process illustrated in Fig. 3.12, where only one sample is acquired at each sampling period, imposes some serious restrictions and requires rather high sampling frequencies. Another approach called the oversampling-based feedback acquisition is described in Section 3.3.

3.2.5 Low-Pass RC Filter as an Anti-alias Filter

The anti-aliasing filter has to remove the frequency content above one half of the sampling frequency ($\Omega/2$) from the analogue input signal. Therefore, it has to be an analogue, continuous-time filter, having the transfer function $W_F(s) = N(s)/D(s)$. For the low-pass filter, and for large frequencies $s = j\omega$, the magnitude of the polynomial $N(s)$ has to be smaller than the value of the polynomial $D(s)$. Therefore, the order of $D(s)$ has to be larger than the order of $N(s)$. With an R-C filter,

$$W_F(s) = \frac{u_{ADC}(s)}{u_{ADC1}(s)} = \frac{1}{1 + RC \cdot s}, \quad W_F(j\omega) = \frac{1}{1 + j\omega RC}. \tag{3.26}$$

Passive anti-aliasing R-C filter is given in Fig. 3.18. The frequency characteristic of the R-C filter is given in Fig. 3.19. At the frequency of the real pole $\omega_p = 1/(RC)$, the amplitude characteristic drops down by 3 dB, that is, the sinusoidal input u_{ADC1} at the frequency ω_p will produce the output signal u_{ADC} with sqrt(2) lower amplitude, and with the corresponding power divided by 2. At frequencies $10 \cdot \omega_p$ and $100 \cdot \omega_p$, the attenuation will be 20 and 40 dB; in other words, the amplitude of the input signal will be 10 times and 100 times lower, respectively.

The current supplied from a grid-side inverter through the LCL filter into the grid has a quasi-sinusoidal waveform with superimposed current ripple, as shown in Figs. 2.4 and 2.12, wherein the ripple amplitude is shown in (2.2). Similar current ripple is obtained in cases where a PWM inverter supplies an ac motor. From (2.2), and assuming that the PWM frequency is 10 kHz, while the series inductances of the LCL filter are equal to 0.1 p.u, the current ripple ΔI exceeds 9% of the rated current (answer to the Question 2.1).

In double-update-rate scheme, explained in Sect. 2.1.5, the sampling period T_S is equal to $T_{PWM}/2$. Since the anti-alias filter must remove all the content above one half of the sampling frequency, this means that the RC filter has to remove all the content at $f_{PWM} = 1/T_{PWM}$ and above f_{PWM} in Fig. 2.10. An RC filter with the real pole $2\pi f_p = \omega_p = 1/(RC) < 2\pi f_{PWM}$ attenuates the current ripple at the PWM frequency $A_T = f_{PWM}/f_p$ times. In other words, if the RC filter is designed for $f_p = 100$ Hz, and the PWM frequency is 10 kHz, the attenuation of the current ripple is going to be $A_T = 100$ times. If the current ripple ΔI reaches 10%, and the

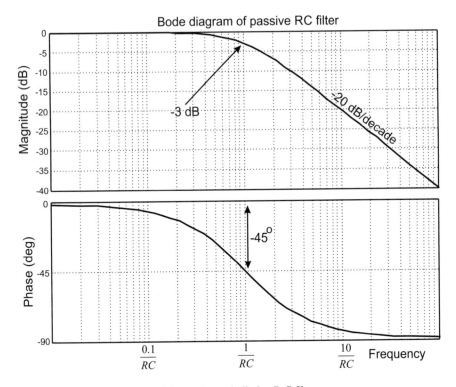

Fig. 3.19 Frequency response of the passive anti-aliasing R-C filter

attenuation A_T is 100, then the residual ripple within the input u_{ADC} of the A/D converter reaches 0.1% of the rated current. Thus, the goal of removing the spectral content above one half of the sampling frequency is not achieved. As a matter of fact, it cannot be achieved with an RC filter. Namely, whatever the cutoff frequency, the attenuation $A_T = f_{PWM}/f_p$ is always limited.

In practical applications, the resolution of the A/D converter is always limited. With the input range of u_{ADC} from 0 V up to $U_{MAX} = 3$ V and with $N = 12$ bit ADC, the smallest detectable change is $\Delta u_Q = U_{MAX}/2^N = 732$ μV. Considering practical control circuits embedded within PWM-controlled power converters, and taking into account the *digital* noise generated by the fast-switching TTL signals within the DSP, the noise within the input u_{ADC} signal can reach Δu_Q. Therefore, the effect of using the A/D converters with $N > 12$ bits is rather limited. Namely, in cases where the environmental noise reaches Δu_Q, the use of 14-bit or 16-bit A/D converters will have only a limited contribution to the overall precision, while the A/D conversion time will be prolonged due to the need to process more bits of the output word.

With the above considerations in mind, the goal in designing the anti-aliasing filter could be reformulated. Since the smallest detectable change of u_{ADC} is Δu_Q, it is sufficient to reduce the amplitude of the spectral content above one half of the sampling frequency below the level of Δu_Q. With $N = 12$, Δu_Q is close to 0.025% of the full range U_{MAX}. Assuming that the range $[0 \ldots U_{MAX}]$ corresponds to change of the measured current from $-2I_{nom}$ up to $+2I_{nom}$, the residual u_{ADC} ripple of 0.1% of the rated current I_{nom} corresponds to Δu_Q. Considering the above design example, the RC filter designed for $f_p = 100$ Hz proves sufficient to suppress the residual ripple down to the level of Δu_Q, also called *the level of one least significant bit*.

Although the RC filter with $f_p = 100$ Hz can suppress the ripple, it also introduces delays in the control loop. The current controllers are expected to reach the closed-loop bandwidth in excess of 1 kHz, and this cannot be achieved with a low-frequency RC filter introduced into the feedback path. Therefore, it is necessary to find other solutions.

3.2.6 Second-Order Low-Pass Anti-alias Filter

With the first-order RC filter, the amplitude characteristic of the filter has the slope of 20 dB/decade. Therefore, in order to achieve sufficient attenuation at the ripple frequency, the cutoff frequency of the filter has to be very low, and it has negative effects on the closed-loop bandwidth of the current controller.

The slope of the amplitude characteristic can be steeper, provided that the filter has more poles. With the low-pass filter with Q poles and with no zeros, the amplitude characteristic at high frequencies reduces with the slope of $Q \cdot 20$ dB/decade. This opens the possibility to obtain the desired attenuation at the PWM frequency with considerably higher cutoff frequency of the filter. While the attenuation of the RC filter is $A_T = f_{PWM}/f_p$, the second-order filter has an attenuation close to $A_T = (f_{PWM}/f_p)^2$, leading to considerably larger values of f_p for the given A_T and f_{PWM}. With larger values of f_p, delays introduced into the feedback line are lower.

One way of designing the second-order filter is the use of two passive RC filters. Series connection increases the equivalent series resistance, while the nature of the passive RC filter makes the implementation of conjugate-complex poles impossible. For that reason, the second-order filters are mostly designed as *active filters*, namely, the filters that make use of operational amplifiers. An example of the second-order anti-aliasing filter organized around an operational amplifier is shown in Fig. 3.20. The transfer function of the filter is given in (3.27):

$$\frac{-u_{ADC}(s)}{u_{ADC1}(s)} = \frac{1}{R_1 R_3 C_1 C_2 \cdot s^2 + \frac{R_1 R_2 + R_1 R_3 + R_2 R_3}{R_2} C_1 \cdot s + \frac{R_3}{R_2}} \qquad (3.27)$$

Polynomial in denominator of (3.27) is of the second order. Design values of resistances and capacitors can be selected to provide either real or conjugate complex poles. At the same time, the ratio between R_2 and R_3 determines the static gains, which can be adjusted to be larger than 1. The amplitude characteristic at high frequencies reduces with the slope of 40 dB/decade. Considering the filter cutoff frequency f_p, attenuation of the second-order filter at the PWM frequency is close to

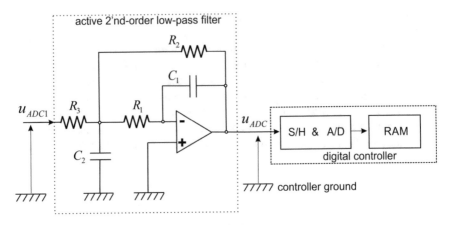

Fig. 3.20 Second-order anti-aliasing filter

$A_T = (f_{PWM}/f_p)^2$. Compared to the attenuation of the passive RC filter ($A_T = f_{PWM}/f_p$), the attenuation of the current ripple is f_{PWM}/f_p larger when using the filter of Fig. 3.20. It is also possible to make third- and fourth-order active filters and to achieve even larger attenuation. At the same time, the use of active components such as the operational amplifiers brings in the consequential offset and noise. Therefore, it is necessary to study further means of the error-free feedback acquisition.

3.2.7   Center-Pulse Sampling

Desired closed-loop bandwidth of current controllers goes beyond 2 kHz both in ac motor drives and in grid-side power converters. Therefore, the anti-alias filters with large time constants and considerable delays have to be avoided. Instead, the suppression of the current ripple can be performed by the *synchronous sampling* technique, illustrated in Fig. 3.21. The figure illustrates *double update rate*, namely, the current samples are acquired at the center of both positive and negative voltage pulses, at instants that coincide with the peaks of the triangular PWM carrier. The waveform of the current ripple is derived for the single-phase representation (Fig. 2.6) of one inverter phase. Small rectangles denoted by *EXE*, placed at the bottom of the figure, represent the intervals where the digital controller executes, processing the newly acquired current sample and calculating the new voltage reference (i.e., the modulation index).

The samples are taken twice in each T_{PWM} period, in the middle of both positive and negative voltage pulses, at instants denoted by t_1 and t_2 in the figure. The waveform of the current ripple in Fig. 3.21 is obtained with assumption that the series resistance is negligible and that the current changes at a constant slope, defined by the supply voltage. Therefore, the zero-crossing of the ripple coincides with the sampling instants, and the samples obtained at the center of the voltage pulses do not

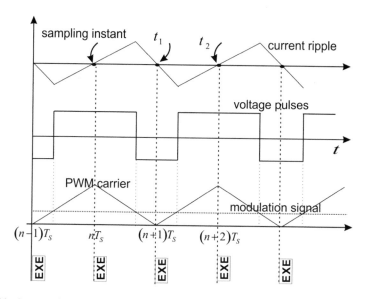

Fig. 3.21 Center-pulse sampling with simplified waveform of the current ripple, obtained with the single-phase representation of the load and with the assumption that the series resistance is negligible

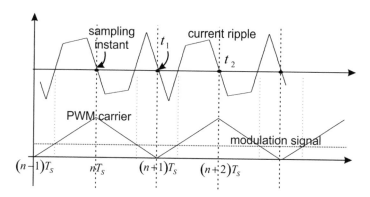

Fig. 3.22 Center-pulse sampling with simplified waveform of the current ripple, obtained with the three-phase representation of the load and with $R = 0$

get affected by the ripple. Thus, it is not necessary to apply the low-pass anti-alias filters with their consequential delays.

In practical three-phase inverters, where the load current depends on the line-to-line voltages, the single-phase representation of Fig. 2.6 does not hold. The current ripple in one phase gets also affected by the other two-phase voltages, and it has a different waveform, shown in Fig. 2.12 and enlarged in Fig. 3.22. As in Fig. 3.21, the waveform of the current ripple is obtained with $R = 0$. At instants where the triangular PWM carrier reaches the maximum and minimum, at the center of the

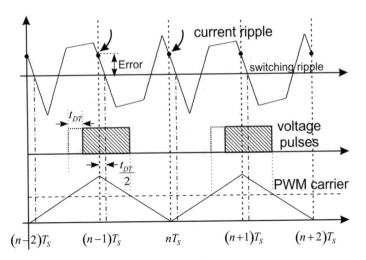

Fig. 3.23 Center-pulse sampling with simplified waveform of the current ripple, obtained with the three-phase representation of the load and with $R = 0$, and with the lockout time t_{DT}

voltage pulses, the current ripple crosses zero, and the acquired samples of the output current do not get affected by the ripple.

Synchronous sampling allows the use of the anti-aliasing filters with considerably larger cutoff frequencies. Yet, it provides the feedback signal from only one sample in each $T_S = T_{PWM}/2$. In cases where the zero-crossing of the ripple does not coincide with the sampling instant due to delays or imperfections, the feedback signal comprises the sampling error. Sampling errors can be introduced due to the lockout time, which is described in Sect. 2.3. The waveform of the current ripple, the voltage pulses, and the PWM carrier shown in Fig. 3.23 are obtained by introducing the necessary lockout time t_{DT}. The most common way of implementing the lockout time is delaying the rising edge of the gating signal for the ongoing switch by t_{DT} while making no delay in passing the falling edge of the gating signal for the offgoing switch. As a consequence, the zero-crossing of the current ripple is delayed by $t_{DT}/2$ due to the lockout time. For this reason, the acquired feedback sample contains an error, shown in Fig. 3.23.

Previous considerations were based on the assumption that the equivalent series resistance R is equal to zero. The equivalent series resistance includes the load resistance but also the output resistance of the inverter and the resistance of the cables. It is small, leading to relatively large L/R time constants that could exceed 10 ms. Even so, the presence of the resistance R changes the constant slope waveform A (Fig. 3.24) into the exponential waveform B. For this reason, the sampled ripple current is not equal to zero ($i_1 = 0$), thus resulting in the sampling error (i_2).

In addition to the lockout time-related errors and the finite-resistance-related errors, the errors in the center-pulse-based feedback acquisition method are also caused by disturbances other than the ripple. With only one sample per T_S period,

Fig. 3.24 The impact of the load resistance on the center-pulse sampling. Simplified waveform of the current ripple is obtained with single-phase representation of the load. The waveform A corresponds to $R = 0$, while the waveform B corresponds to $R > 0$

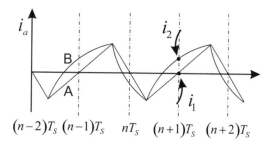

the method is prone to the noise that may come from large dv/dt values of the PWM voltage waveforms. In conjunction with parasitic capacitance of the cables, windings, and inductances, the steep change of the PWM voltage pulses could give a rise to considerable, poorly damped oscillations. These oscillations affect the output current and the phase voltage. Through the direct path and through parasitic couplings, the oscillations affect the signals $u_{ADC}(t)$ and $u_{ADC1}(t)$. In cases where the PWM switching comes several microseconds before the sampling instant, the oscillations retain considerable amplitude, and they may introduce considerable errors. Other noise signals may produce similar effects. In order to reduce the noise sensitivity and reduce delays caused by anti-aliasing filters, it is possible to replace the center-pulse single-sample-based feedback acquisition by the sampling scheme where the feedback is obtained by averaging a larger number of equidistant samples.

3.3   Oversampling-Based Feedback Acquisition

The voltage pulses obtained at the output of the three-phase inverter comprise the frequency components at the PWM frequency f_{PWM} and its multiples (Fig. 2.10). Therefore, the current ripple that comes in consequence has the same frequency components. For this reason, the feedback acquisition system should be capable of removing the signals at the PWM frequency and its multiples. This can be achieved by collecting $N_{OV} = 2^N$ equidistance samples within each PWM period $T_{PWM} = 2T_S$, by calculating the average value of these samples, and by using the obtained one-period average as the feedback signal.

3.3.1   One-PWM-Period Averaging

The output phase current $i_a(t)$ comprises the fundamental frequency component $i_{aF}(t)$ and the superimposed current ripple $\Delta i_a(t)$. The ripple can be approximated by the sum of the frequency components $n \cdot f_{PWM}$, where n is an integer:

$$\Delta i_a(t) = \sum_{n=1}^{+\infty} A_n \cos\left(n \cdot 2\pi f_{\text{PWM}} + \varphi_n\right), \quad f_{\text{PWM}} = \frac{1}{T_{\text{PWM}}} = \frac{1}{2T_S} \qquad (3.28)$$

All the frequency components of $\Delta i_a(t)$ have an integer number (n) of their periods (T_{PWM}/n) within each PWM period. Therefore, their average value, as well as the average value of $\Delta i_a(t)$ over each PWM period T_{PWM}, is equal to zero. Thus, the best way to remove the frequency components of the current ripple from the feedback signal is to calculate an average over each PWM period and use the result as the feedback signal:

$$i_a^{\text{FB}}(nT_S) = \frac{1}{2T_S} \int_{(n-2)T_S}^{nT_S} i_a(t) \cdot dt. \qquad (3.29)$$

In practical implementation, it is difficult to organize analogue integration. Instead, the feedback signal is sampled at the sampling rate $N_{\text{OV}}/2$ times larger than the sampling rate $1/T_S$. The number N_{OV} is usually 16, 32, or 64. The process of acquiring a considerably larger number of equidistant samples is also called the *oversampling*. The number of samples acquired in each PWM period is N_{OV}. The feedback signal i^{FB} is obtained by calculating the average value of the samples acquired in each PWM period.

3.3.2 Oversampling and Averaging

The schedule of the feedback acquisition and the digital current control is synchronized with the PWM carrier. The current controller relations are processed each $T_S = T_{\text{PWM}}/2$. The time required for the controller to execute is indicated by rectangular *EXE* signs in Fig. 3.25. Each *EXE* session is triggered by the interrupt at instants nT_S. In each session, the feedback signal i_n^{FB} is compared to the reference. Based upon the error, the controller calculates the new voltage reference u_n^*. The voltage reference u_n^* becomes available within the interval $[nT_S \ldots (n+1)T_S]$, at instant when the execution of the controller ends. Not being available at the very beginning of the interval, the value u_n^* cannot be used to control the pulse width and the average value of the output voltage within the interval $[nT_S \ldots (n+1) T_S]$. Instead, the value u_n^* is used to control the average voltage on the next interval $[(n+1)T_S \ldots (n+2)T_S]$.

The feedback signal i_n^{FB}, used in calculations triggered by nT_S, is obtained from the samples acquired within the previous PWM period, on the interval $[(n-2)T_S \ldots nT_S]$. With oversampling period $T_{\text{ADC}} = T_{\text{PWM}}/N_{\text{OV}}$, there is a total of N_{OV} equidistant samples. The number of samples in each T_{PWM} is usually $N_{\text{OV}} = 2^N$, and it is often 16, 32, or 64. The feedback signal is calculated from (3.30).

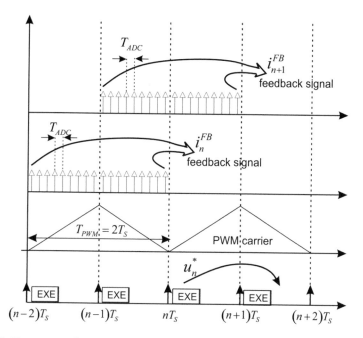

Fig. 3.25 The oversampling scheme which collects $N_{OV} = T_{PWM}/T_{ADC}$ samples in each PWM period and calculates the feedback signal i^{FB} as an average of the samples. Execution of the control interrupts is denoted by EXE. In interrupt session triggered at nT_S, the feedback signal i_n^{FB} is calculated as the average value of $i_a(t)$ over the interval $[(n-2)T_S \ldots nT_S]$

$$i_n^{FB} = \frac{1}{N_{OV}} \sum_{k=0}^{N_{OV}-1} i_a(nT_S - kT_{ADC}). \tag{3.30}$$

3.3.3   Practical Implementation

Assuming that the number of samples collected in each PWM period is $N_{OV} = 32$, and that the PWM frequency is $f_{PWM} = 10$ kHz, then the actual period of sampling is $T_{ADC} = 100\ \mu s/32 = 3.125\ \mu s$. In order to prevent the alias frequency components within the train of samples, the analogue input to the A/D converter must be filtered to suppress the frequency components above one half of the sampling frequency. With $N_{OV} = 32$ and $f_{PWM} = 10$ kHz, one half of the sampling frequency is $f_{OH} = 1/2/$ $3.125\ \mu s = 160$ kHz.

Considerably larger sampling frequency simplifies the design of the anti-aliasing low-pass filter. The task of the filter is suppression of the frequency components at f_{OH} and above f_{OH} down to the level of Δu_Q, which is *one least significant bit* of the

A/D converter. Namely, with a 12-bit A/D converter, the level of Δu_Q corresponds to 1/4096 fraction of the full input range.

In typical application, studied in Questions 3.6 and 3.7, the current ripple at the PWM frequency of $f_{PWM} = 10$ kHz amounts roughly $50 \cdot \Delta u_Q$. Due to the low-pass nature of the load, it is a reasonable assumption that the frequency content at $f_{OH} = 160$ kHz is going to be at least four to five times lower. With the presumed amplitude of $12 \cdot \Delta u_Q$, the frequency components at f_{OH} have to be attenuated below Δu_Q; thus, the required attenuation of the anti-aliasing RC filter is $A_T = (f_{OH}/f_p) = 12$, where $f_p = 1/(RC)/(2\pi)$ is the cutoff frequency of the filter. In the end, $f_p = 13.33$ kHz, the frequency considerably above the desired closed-loop bandwidth of the current controller. Delay of $\tau = RC = 11.9$ μs has a negligible effect on the closed-loop performance. According to considerations in Question 3.8, delay τ limits the closed-loop bandwidth to $f_{BW} = 1/(3\tau)/(2\pi) = 4.444$ kHz. Further improvements can be achieved by selecting a larger N_{OV}.

The oversampling process, the storage of sample trains, and the calculation of their average are facilitated and automated by contemporary digital signal controllers such as TMS320F28335. Device includes a fast 12-bit, 16-channel A/D peripheral unit with declared 80 ns sampling rate. The A/D conversion process can be synchronized with the PWM pulse generator. The oversampling process can be automated, and the internal DMA unit can be programmed to collect the samples and store them in designated areas of RAM memory in an effortless manner, without any involvement of the CPU unit. Described tools can be used to acquire T_{ADC}-spaced samples of the output current and to place them in the internal RAM. In this way, the interrupt-triggered current control routine has to read the train of samples collected during the past PWM period and to calculate the feedback signal as an average value of the past $N_{OV} = T_{PWM}/T_{ADC}$ samples. With the instruction cycle of 6.67 ns, the process completes in less than 300 ns even with $N_{OV} = 32$.

3.3.4 Pulse Transfer Function of the Feedback Subsystem

The average in (3.30) operates on T_{ADC}-spaced samples acquired within one whole PWM period T_{PWM}. Adopting the oversampling period T_{ADC} as the sampling time, assuming that the number of samples collected in each PWM period is $N_{OV} = 32$, and introducing the operator z, wherein the multiplication by z^{-1} designates delay by one sampling period, the pulse transfer function $i^{FB}(z)/i(z)$ is derived in (3.31).

$$i_n^{FB} = \frac{1}{32}\sum_{k=0}^{31} i_a(nT_S - kT_{ADC}) \Rightarrow$$

$$i^{FB}(z) = \frac{1}{32}i_a(z) \cdot \left(1 + z^{-1} + z^{-2} + \cdots + z^{-31}\right) \Rightarrow \tag{3.31}$$

$$\frac{i^{FB}(z)}{i_a(z)} = W_{FIR}(z) = \frac{1}{32}\sum_{k=0}^{31} z^{-k}.$$

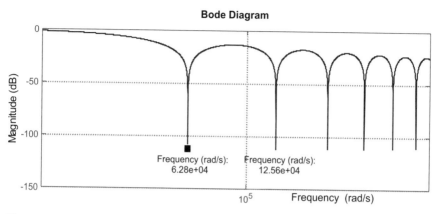

Fig. 3.26 The frequency characteristic of the one-period averaging filter in (3.31). The filter $W_{FIR}(z)$ has an infinite attenuation at frequencies $n \cdot f_{PWM}$. In the plot, the attenuation appears limited to -110 dB due to the finite wordlength used in Bode plot calculations

The frequency characteristic of the filter $W_{FIR}(z)$ is given in Fig. 3.26. The filter has an infinite attenuation at frequencies $n \cdot f_{PWM}$. Within the Bode plot, the attenuation appears limited to -110 dB due to the finite wordlength used in calculations. The Bode plot illustrates the capability of the one-PWM-period averaging method to remove all the PWM-related frequency component from the feedback signals.

In addition to removing the undesired frequency components, the oversampling/ one-PWM-period averaging method also introduces delay into the feedback path. Namely, the feedback signal i^{FB} used at instant nT_S uses the average value of the feedback signal over the past PWM period, thus introducing delay of, approximately, one sampling period T_S. For the purposes of designing the closed-loop control system, it is necessary to derive the model of the delay. The FIR transfer function represents an accurate model of the process. Yet, it uses the oversampling period T_{ADC} as the sampling time. By adopting the model $W_{FIR}(z)$ of (3.31), the process of designing the controller and setting the control parameters has to consider two sampling rates, T_S and T_{ADC}. The formal approach can employ the modified z-transformation. In order to avoid complexity of the modified z-transformation, it is possible to develop single-sampling-rate model of the feedback acquisition system that would correspond to the one in (3.31) in the region below the sampling frequency, which is the region of interest when it comes to the design and parameter setting of the closed-loop current controller.

In Fig. 3.25, the feedback sample i_n^{FB} is calculated within the routine triggered at nT_S as an average value of the current $i_a(t)$ over the past PWM period. Neglecting the ripple and assuming that the rate of change of the current is linear within each sampling period T_S, the feedback i_n^{FB} can be calculated in terms of the values of the actual current at instants $(n-2)T_S$, $(n-1)T_S$, and nT_S, namely, the samples i_{n-2}, i_{n-1}, and i_n. With presumed linear change, the average value on the interval $[(n-2) T_S \ldots (n-1)T_S]$ is equal to $(i_{n-2} + i_{n-1})/2$. Similarly, the average value on $[(n-1) T_S \ldots nT_S]$ is $(i_{n-1} + i_n)/2$. From there, the average value on the past PWM period

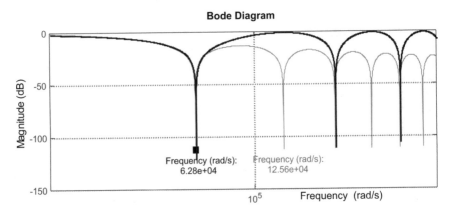

Fig. 3.27 The frequency characteristic of the pulse transfer function of (3.32). In the frequency range below one half of the sampling frequency, the frequency characteristic corresponds to the one in Fig. 3.26. The differences are observed beyond one half of the sampling frequency

$T_{PWM} = 2T_S$ is equal to $(i_{n-2} + 2i_{n-1} + i_n)/4$. The feedback samples i_n^{FB} can be represented by their z-domain image $i^{FB}(z)$, while the samples of the actual current i_n can be represented by $i(z)$. From the above considerations and using the operator z^{-1} to mark the delay of one sampling period, the pulse transfer function $W_{FB}(z)$ of the feedback system becomes

$$W_{FB}(z) = \frac{i^{FB}(z)}{i(z)} = \frac{z^{-2} + 2 \cdot z^{-1} + 1}{4} = \frac{z^2 + 2 \cdot z + 1}{4z^2}. \tag{3.32}$$

The frequency characteristic of the pulse transfer function $W_{FB}(z)$ is shown in Fig. 3.27, along with the frequency characteristic of $W_{FIR}(z)$, included for the comparison and printed in gray. In the frequency range below one half of the sampling frequency $1/(2T_S)$, the two characteristics coincide. At higher frequencies, there are considerable differences between the two. The purpose of the pulse transfer function $W_{FB}(z)$ is to represent the feedback acquisition system within the frequency range of interest for the design of the closed-loop controller, that is, from zero up to one half of the sampling frequency $1/(2T_S)$. Namely, the sampling process prevents higher frequencies, and they do not enter the closed-loop controller. Therefore, the differences between $W_{FB}(z)$ and $W_{FIR}(z)$ at frequencies $f > 1/(2T_S)$ are not relevant as they do not have any impact on the controller design. For that purpose, the proposed one-PWM-period averaging feedback system can be represented by the model $W_{FB}(z)$.

3.3.5 Current Measurement in LCL Filters

In order to remove the PWM ripple injection, grid-side inverters connect to the mains through *LCL* filters, such as the one shown in Fig. 3.28, which represents the single-

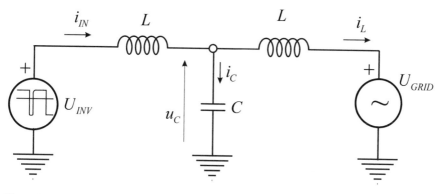

Fig. 3.28 LCL filter with two identical inductances and one parallel capacitor. For the stable operation, it is necessary to introduce the control action proportional to the capacitor current

phase equivalent of the three-phase inverter. The inverter voltage U_{INV} comprises a train of PWM pulses, which generates the current i_{IN}. The current includes the fundamental, line-frequency component and also the current ripple caused by the PWM voltages. The parallel capacitor has a very low impedance at the PWM frequency, and it takes most of the current ripple, leaving only a very small amount of the ripple in i_L that circulates into the mains.

The *LCL* filter is a resonant tank, prone to oscillations. This makes the task of designing the current controller more difficult. In order to take a closer insight into design problems, it is of interest to analyze the impact of the inverter voltage U_{INV} on the output current i_L. The transfer function $Y(s)$ of (3.33) provides relation between $i_L(s)$ and $U_{INV}(s)$, obtained in the case where $U_{GRID} = 0$. The transfer function $Y(s)$ has conjugate complex poles at $s_{1/2} = \pm j \cdot \text{sqrt}(2/L/C)$, and these poles give rise to undamped oscillations. In order to introduce the damping of the poles $s_{1/2}$, it is necessary to measure the capacitor current $i_C(t)$ and to introduce the control action proportional to the capacitor current.

$$Y(s) = \frac{i_L(s)}{U_{INV}(s)} = \frac{1}{L \cdot s} \cdot \frac{1}{2 + LC \cdot s^2}. \tag{3.33}$$

In practical implementation, the measurement of the current i_C has to be performed in very much the same manner as the measurement of the output current, thoroughly described in previous sections. Therefore, the measurement process will include the anti-alias filtering and the delays, such as the one described by $W_{FB}(z)$ of (3.32). The objective of the present discussion is to underline the basic principles of stabilization by means of the capacitor-current feedback. Therefore, the measurement is assumed to be instantaneous, and the analysis is performed in *s*-domain.

In order to introduce the desired damping, the inverter voltage has to be modified. It is necessary to introduce the component proportional to the capacitor current $i_C(t)$. Assuming that the original inverter voltage U_{INV} gets replaced by $U_{INV1} = U_{INV} - ki_C$, the transfer function i_L/U_{INV} changes and assumes the form (3.34)

$$Y_1(s) = \frac{i_L(s)}{U_{INV}(s)} = \frac{1}{L \cdot s} \cdot \frac{1}{2 + k \cdot C \cdot s + LC \cdot s^2}. \tag{3.34}$$

The polynomial in denominator of (3.34) has the roots given in (3.35). By increasing the coefficient k, imaginary part of the roots decreases, while the real part increases, thus increasing the damping. With the feedback coefficient k_{opt}, given in (3.36), both poles of (3.34) are real and equal to $-k/L/2$:

$$s_{1/2} = -\frac{1}{2}\frac{k}{L} \pm j\sqrt{\frac{2}{LC} - \frac{k^2}{4L^2}}. \tag{3.35}$$

$$k_{opt} = \sqrt{\frac{8L}{C}}. \tag{3.36}$$

Question (3.6) The input voltage range of the A/D converter with $N = 12$ bits is $u_{ADC} \in [0 \ldots 3 \text{ V}]$, and it corresponds to the change of the measured current from $-3 \cdot I_{nom} \cdot \text{sqrt}(2)$ up to $+3 \cdot I_{nom} \cdot \text{sqrt}(2)$. The current ripple ΔI is equal to $0.1 \cdot I_{nom}$. The PWM frequency is 10 kHz. Determine the time constant of the RC filter that reduces the residual ripple within u_{ADC} down to the level of one significant bit.

Answer (3.6) Considering the input range and the scaling of the measured current, where the change in u_{ADC} of 3 V corresponds to the change of the measured current by $6 \cdot I_{nom} \cdot \text{sqrt}(2)$, one least significant bit corresponds to voltage change Δu_Q which represents the current of $6 \cdot I_{nom} \cdot \text{sqrt}(2)/2^{12} = 0.0020716 \, I_{nom}$. Thus, desired attenuation of the anti-aliasing filter is $A_T = 0.1/0.0020716 = 48.27$. The sampling frequency is $2/T_{PWM} = 20$ kHz, while the current ripple has the base frequency of 10 kHz. Therefore, the cutoff frequency of the RC filter is $f_{PWM}/A_T = 207$ Hz, which leads to $RC = 768 \, \mu s$.

Question (3.7) For the A/D converter and the current-sensing system of Question 3.6, and for the same amplitude of the ripple current, it is necessary to design the filter of Fig. 3.20 and select the resistances and capacitances so that it has two identical real poles. Desired static gain of the filter is equal to one.

Answer (3.7) As in the previous example, one least significant bit corresponds to voltage change $\Delta u_Q = 0.0020716 \, I_{nom}$, while the desired attenuation of the anti-aliasing filter at $f_{PWM} = 10$ kHz is $A_T = 0.1/0.0020716 = 48.27$. Considering the filter cutoff frequency f_p, attenuation of the second-order filter with two identical real poles at the PWM frequency is close to $A_T = (f_{PWM}/f_p)^2$. Thus, the cutoff frequency of the filter is $f_p = f_{PWM}/\text{sqrt}(A_T) = 1440$ Hz, namely, $\omega_p = 9043$ rad/s. In order to obtain the unity gain, it is necessary to set $R_2 = R_3$. In order to obtain two real poles, the characteristic polynomial in denominator of (3.27) has to be $f(s) = (1 + s/\omega_p)^2$, that is,

$$f(s) = R_1 R_2 C_1 C_2 \cdot s^2 + (2R_1 + R_2)C_1 \cdot s + 1 = \frac{1}{\omega_p^2}s^2 + \frac{2}{\omega_p}s + 1.$$

The remaining degree of freedom can be used to set $R_1 = R_2 = R = 10$ kΩ, the value which loads the operational amplifier output with reasonably low current, yet resulting in sufficiently low equivalent impedances, so as to suppress the impact of the offset currents and the noise currents. From $3RC_1 = 2/\omega_p$, $C_1 = 7.4$ nF. From $R_1R_2C_1C_2 = 1/\omega_p^2$, $C_2 = 16.58$ nF. Considering the available standard values of capacitors, one would select $C_1 = 6.8$ and $C_2 = 15$ nF.

Question (3.8) Consider a simplified analogue current controller with the load model $W_P(s) = 1/L/s$, with the PI current controller and the voltage actuator modeled by $W_{CA}(s) = p + i/s$, and with the delay within the feedback line modeled by $W_D(s) = 1/(1 + s\tau)$. In absence of any other information and using reasonable simplifications, find an estimate of the available closed-loop bandwidth in terms of the delay τ.

Answer (3.8) The closed-loop system comprises $W_P(s)$, $W_{CA}(s)$, and $W_D(s)$. The characteristic polynomial is found from $1 + W_P(s)W_{CA}(s)W_D(s)$, and it reads $f(s) = s^3 + s^2/\tau + s \cdot p/L/\tau + i/L/\tau$. The gain setting and the pole placement of this third-degree polynomial is rather involved. Yet, the question allows the use of reasonable simplifications. Assuming that the closed-loop system has a triple real pole p_1, the characteristic polynomial becomes $f(s) = (s + p_1)^3$. According to Vieta's formulas, $3p_1 = 1/\tau$. Thus, the dominant poles are placed at $p_1 = 1/(3\tau)$, and they determine the closed-loop bandwidth. As an example, delay of $RC = 768$ μs limits the closed-loop bandwidth to, roughly, $f_{BW} = p_1/2/\pi = 71$ Hz.

Question (3.9) The current feedback is obtained by center-pulse sampling, as illustrated in Fig. 3.24. The load inductance is $L = 10$ mH, while the load resistance is $R = 1$ Ω. The finite value of R changes the current-ripple waveform into the form B of Fig. 3.24, thus introducing the sampling errors. Is it possible to design the RC anti-aliasing filter that would bring the zero-crossing of the ripple back to the sampling instant, that is, to the center of the voltage pulses?

Answer (3.9) The center-pulse sampling is based on the ideal load where the equivalent resistance R is equal to zero. Ideally, the waveform of the current ripple (A in Fig. 3.24) is obtained by passing the input voltage pulses through the load transfer function $1/(Ls)$. In practical case (B in Fig. 3.24), the load transfer function is $1/(R + Ls)$. Therefore, the waveform B can be obtained by passing the waveform A through the transfer function $(Ls)/(R + Ls)$. As a consequence, the frequency component at $s = j\omega$ is phase shifted by an advance of atan($R/L/\omega$). With a passive RC filter with $\tau = RC$, the low-pass transfer function $1/(1 + s\tau)$ produces a phase lag of atan($\omega\tau$). For the specific frequency ω, it is possible to select $\tau = RC$ that would compensate the phase advance introduced by the finite load resistance. Yet, the voltage pulses and the current ripple are periodic signal with multitude of frequency components. Even if the value of τ is adjusted to bring the zero-crossing of the ripple to the sampling instant for the given pulse width, the zero-crossing would drift away for other pulse widths.

Question (3.10) The PWM frequency of the system is 10 kHz, while the oversampling A/D converter acquires the samples with the frequency of 640 kHz. In the frequency range above one half of the oversampling frequency $f_{OH} = 320$ kHz, the frequency content within the input signal does not exceed $20 \cdot \Delta u_Q$, where Δu_Q is the signal change that corresponds to one least significant bit of the A/D converter. It is necessary to design the anti-aliasing RC filter. The series resistance should not exceed 1 K.

Answer (3.10) The frequency components at f_{OH} and beyond f_{OH} should be suppressed below the level of Δu_Q. Desired attenuation of the anti-aliasing RC filter is $A_T = (f_{OH}/f_p) = 20$. Therefore, $f_p = 1/(RC)/(2\pi) = f_{OH}/20 = 16$ kHz. The time constant of the filter is $\tau = RC = 9.95$ µs. With $R = 1$ kΩ, the parallel capacitor is $C = 9.95 \approx 10$ nF.

Question (3.11) The LCL filter has the series inductances of $L = 10$ mH each and the parallel capacitor of $C = 10$ µF. It is possible to measure the capacitor current and to change the inverter voltage by the amount $-ki_C$. What is the minimum gain k that results in real poles of the transfer function $Y(s) = i_L(s)/U_{INV}(s)$? What is the frequency of those poles?

Answer (3.11) As the gain k in (3.35) increases, the imaginary parts of the two conjugate complex poles decrease. When the gain reaches k_{opt} of (3.36), the imaginary part reaches zero, and the two poles are real and equal. For larger values of k, the two poles detach. Thus, desired minimum value of k is $k_{opt} =$ sqrt(8/L/C) = 89.44. Corresponding poles are $s_{1/2} = -k/L/2 = -4472$ rad/s, which corresponds to 712 Hz.

Chapter 4
Introduction to Current Control

This chapter introduces the basics of three-phase digital current controllers. Their practical implementation is digital, and it involves the PWM actuator and the feedback acquisition systems described in the previous chapter. The analysis and design of discrete-time current controllers rely on z-transform, and it takes into consideration all the feedback acquisition, computation, and PWM delays. In order to facilitate the introduction of some basic principles of the current control, this chapter does not take into account the discrete nature of the controller, nor does it consider the transport delays. Instead, the analysis is performed in s-domain.

The current controllers are used in grid-side inverters but also in ac drives. In both applications, the model of the load is similar, and it comprises the series impedance and a back-electromotive force (Fig. 4.1). The line voltage of ac grid corresponds to the back-electromotive force of ac machines, while the resistance and the equivalent series inductance of the ac machine windings correspond to the series resistance and inductance placed between the grid-side inverter terminals and the connection to the ac grid.

The sum of the output currents within the three-phase model is equal to zero. Therefore, $i_C = -i_A - i_B$ is not an independent variable. To improve clarity, the variables of the three-phase model are transformed into stationary α–β coordinate frame. Thus, the currents i_A, i_B, and i_C are represented by their projections on the orthogonal axes of the α–β frame. The currents i_α and i_β constitute the current vector $i_{\alpha\beta}$ the α–β frame. Analysis and design of the three-phase current controllers are simplified by representing the current and voltage vectors by complex numbers, such as $i_{\alpha\beta} = i_\alpha + ji_\beta$.

In both the grid-side inverters and ac drives, the steady-state output currents are ac currents. Their frequency is either the line frequency of the ac grid, or the frequency that corresponds to the speed of the revolving magnetic field within ac machines. Due to finite gains and a finite closed-loop bandwidth of the α–β frame current controller, the operation with ac current references introduces the phase and amplitude errors. In order to obtain an error-free current control of line-frequency

© Springer International Publishing AG, part of Springer Nature 2018
S. N. Vukosavic, *Grid-Side Converters Control and Design*, Power Electronics and Power Systems, https://doi.org/10.1007/978-3-319-73278-7_4

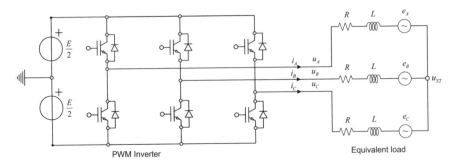

Fig. 4.1 In both grid-side inverters and ac drives, the three-phase inverter is connected to the three-phase load which comprises the series impedance and back-electromotive force

ac currents, it is necessary to transform the voltage and current vectors in d-q coordinate frame that revolves in synchronism with the grid voltages and currents. In steady state, the d-q components of the voltages and currents (i.e., projections of relevant vectors on the axes of synchronously revolving frame) are constant. Therefore, a simple PI controller in d-q frame will provide an error-free operation in the steady state. The basic characteristics of the PI current controller in d-q coordinate frame are explored and explained. The current controllers in d-q frame are often called the *synchronous frame current controllers*.

The current components in d-q frame are i_d and i_q. In grid-side inverters with the line voltage aligned with q-axis, the current i_q determines the active power P, while the current i_d determines the reactive power Q. In ac drives where the d-axis is aligned with the rotor flux, the current i_q determines the electromagnetic torque, while the current i_d affects the amplitude of the flux. In both cases, it is not desirable to have any coupling between the d-axis and the q-axis. With simple PI controller in d-q frame, the transients in one of the axes produce considerable effect on the variables of the other axis and vice versa. In order to prevent such undesirable coupling, it is necessary to use the principles of the internal model control (IMC) in order to design the decoupling controller. The application of the IMC concept is introduced and explained in s-domain.

The d-q frame current controller provides an error-free control of the line-frequency currents. At the same time, the implementation of the d-q frame controller requires several coordinate transformations, including the revolving Park transformation. At times where the computing power of digital signal controller was limited, this circumstance was a serious disadvantage of the d-q frame controllers. Moreover, not even the d-q frame controller which revolves synchronously with the grid can eliminate the inverse components, which appear in nonsymmetrical three-phase systems. Similarly, it cannot suppress the low-frequency line harmonics. The current controllers in stationary α–β frame do not use the Park transformation. In their integrators, the factor $1/s$ can be replaced by factors $s/(s^2 + \omega^2)$, thus resulting in *resonant* controllers, which are capable to remove any error at the frequency ω. One resonant controller can use several factors of the form $s/(s^2 + \omega^2)$, each having a different frequency ω, contributing to elimination of several line harmonics. Steady-state and transient properties of resonant controllers are studied and explored using a simplified model in *s*-domain.

The current controllers are expected to suppress the errors and to bring the output current to the desired reference waveform. The errors can be caused by the change of the input reference. At the same time, the errors can appear due to the changes in line voltage or due to the changes in the electromotive force. Such changes represent the voltage disturbance. The change of the current, caused by the voltage disturbance, should be as low as possible, preferably zero. Voltage disturbance rejection capability is characterized by the output admittance of the current controller, $Y(j\omega) = \Delta i$ $(j\omega)/\Delta u(j\omega)$, where Δi is the current error while Δu is the voltage disturbance. The value $Y(j\omega)$ defines the current error that comes as a consequence of the voltage disturbance at frequency ω. It is desirable to obtain the values $Y(j\omega)$ as low as possible over the whole frequency range where the voltage disturbances could take place. The value of $Y(j\omega)$ can be reduced by using additional control action called the *active resistance feedback*. The impact of the active resistance feedback is studied in *s*-domain.

Introductory considerations outlined in this chapter are focused on explaining the main structures and features of the current controller. In order to focus on essentials, all the developments are simplified, and all the analyses were done in *s*-domain. In order to apply the knowledge and design actual current controllers, the reader must study the subsequent chapters which discuss the digital implementation of the main structures and which establish the parameter setting procedures.

4.1 The Model of the Load

In Fig. 4.1, the three-phase inverter supplies the three-phase load, represented by star-connected phases, each of them having a series resistance, a series inductance, and a back-electromotive force. The voltages of the ac grid correspond to the back-electromotive force of ac machines, while the resistance and the equivalent series inductance of the ac machine windings correspond to the series resistance and inductance placed between the grid-side inverter terminals and the connection to the ac grid. The sum of the three currents $(i_A + i_B + i_C)$ is equal to zero. Therefore, there are only two independent currents in the system. For this reason, it is convenient to represent the three currents by the currents i_α and i_β of the α–β stationary frame, which make up the current vector $\underline{i}_{\alpha\beta}$.

4.1.1 The Three-Phase Load

The load of Fig. 4.1 has the star connection voltage equal to u_{ST}. The voltage balance equations are given in (4.1). The sum of the three currents is equal to zero,

$i_A + i_B + i_C = 0$. Therefore, the sum of the three equations of (4.1) results in (4.2). Assuming that the sum of the back-electromotive forces is equal to zero, the voltage at the star connection is given in (4.3):

$$u_A - Ri_A - L\frac{di_A}{dt} - e_A = u_{ST},$$
$$u_B - Ri_B - L\frac{di_B}{dt} - e_B = u_{ST},$$ (4.1)
$$u_C - Ri_C - L\frac{di_C}{dt} - e_C = u_{ST}.$$

$$u_A + u_B + u_C - (e_A + e_B + e_C) = 3 \cdot u_{ST}.$$ (4.2)

$$u_{ST} = \frac{u_A + u_B + u_C}{3} - \frac{e_A + e_B + e_C}{3} = \frac{u_A + u_B + u_C}{3}.$$ (4.3)

In cases where the sum of the back-electromotive forces is not equal to zero, it is possible to introduce $e_{ST} = (e_A + e_B + e_C)/3$ and express the back-electromotive forces in terms of their average value e_{ST} and the variables e_{A1}, e_{B1}, and e_{C1} (4.4), where $e_{A1} + e_{B1} + e_{C1} = 0$. Even in cases where e_{ST} is not equal to zero, it does not affect the three-phase currents. Instead, it alters the voltage at the star connection. By introducing $u_{ST1} = u_{ST} - e_{ST}$, the voltage balance equations assume the form (4.5):

$$e_{ST} = \frac{e_A + e_B + e_C}{3},$$ (4.4)
$$e_{A1} = e_A - e_{ST}, \quad e_{B1} = e_B - e_{ST}, \quad e_{C1} = e_C - e_{ST}.$$

$$u_A - u_{ST1} = Ri_A + L\frac{di_A}{dt} + e_{A1},$$
$$u_B - u_{ST1} = Ri_B + L\frac{di_B}{dt} + e_{B1},$$ (4.5)
$$u_C - u_{ST1} = Ri_C + L\frac{di_C}{dt} + e_{C1}.$$

Thus, the model of the load can be represented by the voltage balance equations (4.5), where both the sum of the currents and the sum of the back-electromotive forces are equal to zero, while the voltages supplied to each phase are u_A-u_{ST1}, u_B-u_{ST1}, and u_C-u_{ST1}.

4.1.2 The Model of the Load in α–β Coordinate Frame

The model of the load can be simplified by transforming the three-phase quantities into stationary α–β coordinate frame, shown in Fig. 4.2. In Fig. 4.2a, each phase is assigned the unit vectors a_0, b_0, and c_0, and these vectors are displaced by $2\pi/3$. The resulting current vector $i_{\alpha\beta}$ is obtained by summing the three current vectors. In Fig. 4.2b, the same vector $i_{\alpha\beta}$ is represented in α–β frame, in terms of the components

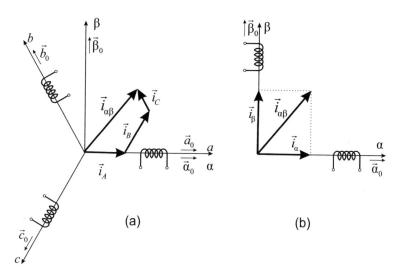

Fig. 4.2 (a) The phases a, b, and c are given the unit vectors a_0, b_0, and c_0, spatially displaced by $2\pi/3$. Each of the currents i_A, i_B, and i_C is represented by the corresponding vector. The vector $i_{\alpha\beta}$ is obtained by summing the three current vectors. (b) The vector $i_{\alpha\beta}$ can be represented in α–β frame, in terms of the components i_α and i_β, which are the projections of the vector $i_{\alpha\beta}$ on the axes α and β

i_α and i_β, and these components are the projections of the current vector on the orthogonal axes of the α–β frame.

Both the A-B-C and α–β components have to create the same current vector. Therefore, relation between the two is defined by Clarke's three-phase to two-phase transformation (4.6). The inverse transformation from α–β to A-B-C components is given in (4.7). The transformations include the leading coefficient K_{23}, which establishes the ratio between the quantities in two coordinate systems. With the leading coefficient $K_{23} = 2/3$, the currents i_A and i_α are equal (4.7). Notice in (4.6) that the average value of the three-phase currents (i_{ST}) is equal to zero:

$$
\begin{bmatrix} i_\alpha \\ i_\beta \\ i_{ST} \end{bmatrix} = K_{23} \begin{bmatrix} 1 & -1/2 & -1/2 \\ 0 & +\sqrt{3}/2 & -\sqrt{3}/2 \\ 1/3 & 1/3 & 1/3 \end{bmatrix} \cdot \begin{bmatrix} i_A \\ i_B \\ i_C \end{bmatrix}. \tag{4.6}
$$

$$
i_A = \frac{2}{3K_{23}} i_\alpha, \quad i_B = \frac{1}{K_{23}\sqrt{3}} i_\beta - \frac{1}{3K_{23}} i_\alpha, \quad i_C = -i_A - i_B. \tag{4.7}
$$

The currents i_α and i_β are obtained in (4.6) from the three-phase currents. Using the same expression, it is possible to transform the voltages u_A, u_B, and u_C into their α–β counterparts. The same holds for the back-electromotive forces; the values e_A, e_B, and e_C can be represented by e_α and e_β. The average values of the voltages u_{ST} (4.3) and the back-electromotive forces e_{ST} (4.4) can alter the voltage at the star connection, but they do not affect the currents. Therefore, the model of the load in the stationary α–β frame is

$$u_\alpha = Ri_\alpha + L\frac{di_\alpha}{dt} + e_\alpha,$$
$$u_\beta = Ri_\beta + L\frac{di_\beta}{dt} + e_\beta. \qquad (4.8)$$

Analysis and design of the three-phase current controllers can be further simplified by representing the current and voltage vectors by complex numbers, such as $i_{\alpha\beta} = i_\alpha + ji_\beta$ and $\underline{u}_{\alpha\beta} = u_\alpha + ju_\beta$. This change in notation consists in replacing the unit vectors α_0 and β_0 by 1 and j. If the second voltage balance equation of (4.8) is multiplied by j and added to the first equation, one obtains the voltage balance equation with complex representation of all the vectors, namely, with $\underline{u}_{\alpha\beta}, \underline{i}_{\alpha\beta}$, and $\underline{e}_{\alpha\beta}$:

$$\underline{u}_{\alpha\beta} = R\underline{i}_{\alpha\beta} + L\frac{d\underline{i}_{\alpha\beta}}{dt} + \underline{e}_{\alpha\beta}. \qquad (4.9)$$

The time-domain Eq. (4.9) can be transformed in s-domain by using the Laplace transform:

$$\underline{i}_{\alpha\beta}(s) = \frac{\underline{u}_{\alpha\beta} - \underline{e}_{\alpha\beta}}{R + L \cdot s}. \qquad (4.10)$$

In (4.10), the voltage $\underline{u}_{\alpha\beta}$ is the input variable that acts as the driving force, the current $\underline{i}_{\alpha\beta}$ is the output of the system, while the back-electromotive force is the voltage disturbance. The task of the current controller is keeping the current equal to the reference in the presence of the disturbances.

4.1.3   The Model of the Load in d-q Frame

The steady-state output currents in both grid-side inverters and ac drives are the ac currents. Their frequency is either the line frequency of the ac grid or the frequency that is related to the speed of the revolving magnetic field in ac machines. Thus, the task of the stationary-frame current controller is to suppress the errors in tracking the ac current references. In cases where the steady-state current references are constant, it is much easier to suppress the current errors. When the voltage and current vectors are transformed from the steady-state α–β frame into d-q coordinate frame which revolves synchronously with the grid voltages (or the revolving field in ac machines), the relevant steady-state projection of the vectors on d- and q-axes is constant. Thus, the d-q components of the relevant voltages and currents are constant, and the control tasks are considerably easier to complete.

In Fig. 4.3, the α–β coordinate frame is stationary. The current vector $i_{\alpha\beta}$ is defined by the projections i_α and i_β of the vector on the corresponding axes. The d-q coordinate frame revolves at the speed ω_e. The speed ω_e is equal to the angular

Fig. 4.3 Transformation of
the output current from the
stationary, α–β coordinate
frame into the
synchronously revolving
d-q frame

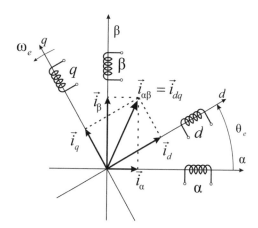

frequency of the grid. In ac drives, it corresponds to the speed of the revolving
magnetic field. The angle between the α-axis and the d-axis is θ_e. At any given
instant, the angle θ_e is defined by (4.11). If the d-q frame revolves in synchronism
with the grid voltages, then it is also synchronous with the current vector. For this
reason, projections of the current vector on axes d and q are constant in the steady-
state operation.

$$\theta_e(t) = \theta_e(0) + \int_0^t \omega_e(\tau) \cdot d\tau \qquad (4.11)$$

Both α–β and d-q components of the current have to provide the same vector in
Fig. 4.3. Relation between the components i_α and i_β and the components i_d and i_q can
be obtained by calculating the projection of the current i_α on the axis d, which is
equal to $i_\alpha \cos(\theta_e)$, and then the projection of the current i_α on the axis q, which is
equal to $-i_\alpha \sin(\theta_e)$. Similarly, the projection of the current i_β on the axis d is equal to
$i_\beta \sin(\theta_e)$, while the projection of the current i_β on the axis q is $i_\beta \cos(\theta_e)$. The above
considerations are summarized in (4.12), which introduces the revolving transfor-
mation of the variables from the stationary frame into the synchronously rotating
frame. This transformation is called the Park transformation:

$$\begin{bmatrix} i_d \\ i_q \end{bmatrix} = \begin{bmatrix} \cos\theta_e & \sin\theta_e \\ -\sin\theta_e & \cos\theta_e \end{bmatrix} \cdot \begin{bmatrix} i_\alpha \\ i_\beta \end{bmatrix} = \underline{T} \cdot \begin{bmatrix} i_\alpha \\ i_\beta \end{bmatrix} \qquad (4.12)$$

The expression (4.12) can be used to transform $\underline{u}_{\alpha\beta}$, $\underline{i}_{\alpha\beta}$, and $\underline{e}_{\alpha\beta}$ and to obtain the
d-q frame variables $\underline{u}_{dq}$, $\underline{i}_{dq}$, and $\underline{e}_{dq}$. Instead of using the unit vectors d_0 and q_0 to
represent the variables in d-q frame as vectors, it is possible to use the d and q axes as
the real and the imaginary axis and to represent the vectors as complex numbers,
such as $\underline{i}_{dq} = i_d + ji_q$ and $\underline{u}_{dq} = u_d + ju_q$. The change in notation consists in
replacing the unit vector d_0 by 1 (real axis) and replacing the unit vector q_0 by j

(imaginary axis). Relying on this notation, the transformation of the variables $\underline{i}_{\alpha\beta}$ $= i_\alpha + ji_\beta$, $\underline{e}_{\alpha\beta} = e_\alpha + je_\beta$ and $\underline{u}_{\alpha\beta} = u_\alpha + ju_\beta$ into the d-q frame can be achieved by multiplying their α-β representation by $\exp(-j\theta_e) = \cos(\theta_e) - j\sin(\theta_e)$:

$$
\begin{aligned}
\underline{i}_{dq} &= i_d + ji_q = e^{-j\theta_e} \cdot \underline{i}_{\alpha\beta} = (\cos\theta_e - j\sin\theta_e) \cdot (i_\alpha + ji_\beta), \\
\underline{u}_{dq} &= e^{-j\theta_e} \cdot \underline{u}_{\alpha\beta}, \quad \underline{e}_{dq} = e^{-j\theta_e} \cdot \underline{e}_{\alpha\beta}.
\end{aligned}
\tag{4.13}
$$

The inverse transformation from the d-q synchronous frame into the α-β stationary frame is performed by multiplying the dq variables by $\exp(+j\theta_e) = \cos(\theta_e) + j\sin(\theta_e)$. Thus,

$$
\underline{i}_{\alpha\beta} = e^{+j\theta_e} \cdot \underline{i}_{dq}, \quad \underline{u}_{\alpha\beta} = e^{+j\theta_e} \cdot \underline{u}_{dq}, \quad \underline{e}_{\alpha\beta} = e^{+j\theta_e} \cdot \underline{e}_{dq}.
\tag{4.14}
$$

The α-β variables from (4.14) can be introduced in (4.9) to obtain

$$
e^{+j\theta_e} \cdot \underline{u}_{dq} = R\underline{i}_{dq} \cdot e^{+j\theta_e} + L\frac{d\left(e^{+j\theta_e} \cdot \underline{i}_{dq}\right)}{dt} + \underline{e}_{dq} \cdot e^{+j\theta_e}.
\tag{4.15}
$$

Resolving (4.15) and dividing the expression by $\exp.(+j\theta_e)$, one obtains the model of the load in d-q coordinate frame:

$$
\underline{u}_{dq} = R\underline{i}_{dq} + L\frac{d\underline{i}_{dq}}{dt} + j\omega_e L\underline{i}_{dq} + \underline{e}_{dq}.
\tag{4.16}
$$

The time-domain Eq (4.16) can be transformed in s-domain by using the Laplace transform:

$$
\underline{i}_{dq}(s) = \frac{\underline{u}_{dq} - \underline{e}_{dq}}{R + L \cdot s + j \cdot L \cdot \omega_e} = W_L(s) \cdot \left(\underline{u}_{dq} - \underline{e}_{dq}\right).
\tag{4.17}
$$

The transfer function $W_L(s)$ in (4.17) represents the model of the load in synchronous, d-q coordinate frame.

4.2   The PI Current Controllers

The model of the load, developed in the previous sections, is a first-order system. Therefore, it is reasonable to attempt the use of the current controller which has proportional and integral actions. When the PI controller discriminates the error $\Delta i_{\alpha\beta}$ in α-β frame, and it calculates the desired voltage command in the same α-β frame, it is called the PI current controller in the stationary frame. In cases where the PI controller discriminates the error Δi_{dq} in d-q frame, and it calculates the desired voltage command in the same d-q frame, it is called the PI current controller in the synchronous frame, of the d-q current controller. It is of interest to study the salient features of both controllers.

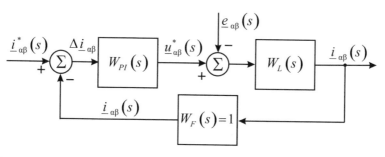

Fig. 4.4 Block diagram of the stationary-frame PI controller

4.2.1 The PI Controller in α–β Frame

The block diagram of the PI controller, implemented in the stationary α–β coordinate frame, is given in Fig. 4.4. The diagram is based on several approximations. It is assumed that the feedback acquisition system has no delay and that the feedback signal represents the actual output current without errors and with no delays. At the same time, it is assumed that the voltage actuator (i.e., the PWM inverter) is ideal, namely, that desired voltage is supplied to the load with no delay, equal to the voltage command which is provided by the current controller.

From (4.10), the transfer function of the load is $W_L(s) = 1/(R + L\,s)$. The transfer function of the PI current controller is $W_{PI}(s) = K_P + K_I/s$. Notice in Fig. 4.4 that the input to the current controller is a complex quantity $\Delta i_{\alpha\beta} = \Delta i_\alpha + j\Delta i_\beta$, which includes two current errors packed within the same complex number. The same holds for the output of the current controller, a complex number which provides the voltage command for the α-axis in its real part and the voltage command for the β-axis in its imaginary part. Since the transfer function $W_{PI}(s)$ of the current controller does not have an imaginary part, the current controller provides the voltage reference where u^*_α depends on the current error Δi_α, and it does not get affected by Δi_β, while the voltage reference u^*_β depends on the current error Δi_β, and it does not get affected by Δi_α.

The closed-loop transfer function of the system in Fig. 4.4 is

$$W_{CL}(s) = \frac{i_{\alpha\beta}(s)}{i^*_{\alpha\beta}(s)}\bigg|_{e_{\alpha\beta}=0} = \frac{W_L(s) \cdot W_{PI}(s)}{1 + W_L(s) \cdot W_{PI}(s)}$$

$$= \frac{1 + s\dfrac{K_P}{K_I}}{1 + s\dfrac{R + K_P}{K_I} + s^2\dfrac{L}{K_I}}. \tag{4.18}$$

In cases where the current reference is a Heaviside step function $\underline{I}^*/s$, the steady-state value of the output is

$$i_{\alpha\beta}(\infty) = \lim_{s \to 0} \left[s \cdot W_{CL}(s) \cdot \frac{I^*}{s} \right] = \frac{\left(1 + 0\frac{K_P}{K_I}\right) \cdot I^*}{1 + 0\frac{R+K_P}{K_I} + 0^2 \frac{L}{K_I}} = I^*. \tag{4.19}$$

Thus, for the Heaviside step as the current reference, the stationary-frame PI controller operates with no error in the steady state. Yet, steady-state currents in grid-side inverters are the line-frequency ac currents. For the given frequency $s = j\omega$, the closed-loop transfer function W_{CL} of (4.18) becomes

$$W_{CL}(j\omega) = \frac{1 + j\omega\frac{K_P}{K_I}}{1 - \omega^2 \frac{L}{K_I} + j\omega \frac{R+K_P}{K_I}}. \tag{4.20}$$

Ideally, with an infinitely large K_I gain, $W_{CL}(j\omega)$ gets equal to one, and the output current corresponds to the reference. In practical implementation, the range of applicable gains is limited by the feedback acquisition delays, the PWM actuator delays, and the noise and unmodeled dynamics. Therefore, the closed-loop transfer function (4.20) gives rise to amplitude and phase errors. As a consequence, the steady-state output currents will not have the desired phase and amplitude. In grid-side inverters, this means that the active and reactive power injected into the grid would not correspond to the desired values. In ac drives, the electromagnetic torque and flux will not coincide with the reference values. In order to avoid the steady-state errors of ac current controllers, it is necessary to transform the PI current controller into the synchronous coordinate frame, where the steady-state values of the current components i_d and i_q are constant.

4.2.2 The PI Controller in d-q Frame

The applicable range of the feedback gains is limited, and they result in a finite closed-loop bandwidth of the PI current controller. Therefore, the operation with ac current references introduces the phase and amplitude errors. For an error-free operation of the current control which injects the line-frequency ac currents into the grid, it is necessary to transform the voltage and current vectors in d-q coordinate frame. The d-q components of the voltages and currents are the projections of relevant vectors on the axes of synchronously revolving frame. In steady state, these values are constant. Therefore, a simple PI controller in d-q frame will provide an error-free operation in the steady state.

The implementation of the d-q frame PI controller is illustrated in Fig. 4.5. Although the implementation is discrete-time and digital, this section presents a simplified analysis of the d-q frame PI controller in s-domain. The feedback acquisition process includes the anti-aliasing filtering, oversampling, averaging, and scaling, and it turns the phase currents into digital form. Within the block denoted by *Clarke*, the currents are transformed in α–β frame. Further on, the *Park* transformation calculates the feedback currents in d-q frame. The current error Δi_{dq} is

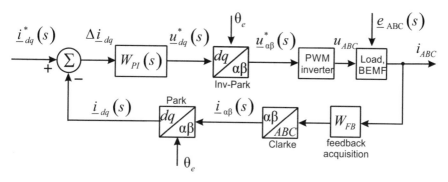

Fig. 4.5 Coordinate transformations of the synchronous d-q frame PI controller

brought to the PI controller, which calculates the voltage reference in d-q frame. The
latter is transformed into α–β frame by the inverse Park transformation. The PWM
inverter is the voltage actuator which supplies desired voltages across the terminals
of the load. The implementation of d-q frame current controllers is numerically
intensive due to several coordinate transformations. At times when fast digital signal
controllers were not available, these calculations were difficult to implement.

Neglecting the delays and assuming that both the PWM actuator and the feedback
acquisition processes are ideal, the block diagram of the PI current controller in d-q
coordinate frame is obtained in Fig. 4.6. The model of the load $W_L(s)$ is given in
(4.17). The closed-loop transfer function of the system of Fig. 4.6 is given in (4.21).

$$
W_{CL}(s) = \frac{i_{dq}(s)}{i_{dq}^*(s)} \bigg|_{e_{dq}=0} = \frac{W_L(s) \cdot W_{PI}(s)}{1 + W_L(s) \cdot W_{PI}(s)}
$$

$$
= \frac{1 + s\dfrac{K_P}{K_I}}{1 + s\dfrac{R + K_P + j\omega_e L}{K_I} + s^2 \dfrac{L}{K_I}} \tag{4.21}
$$

$$
= W_{CLR}(s) + jW_{CLI}(s).
$$

In steady state, the current references in d-q frame are constant. Assuming that the
current reference is a Heaviside step, the current error $\Delta i_{dq}(\infty)$ is equal to zero
(4.22). Constant d-q currents correspond to ac currents in the stationary α–β frame
and the original ABC domain. Thus, the d-q frame PI controller is capable of
injecting ac currents into the grid without any phase or magnitude error.

$$
\Delta i_{dq}(\infty) = \lim_{s \to 0} \left[s \cdot \left(\frac{I^*}{s} - i_{dq}(s) \right) \right]
$$

$$
= \lim_{s \to 0} \left[s \cdot (1 - W_{CL}(s)) \cdot \frac{I^*}{s} \right] \tag{4.22}
$$

$$
= \lim_{s \to 0} \left[s \cdot \frac{s(R + j\omega_e L) + s^2 L}{K_I + s(R + K_P + j\omega_e L) + s^2 L} \cdot \frac{I^*}{s} \right]
$$

$$
= 0.
$$

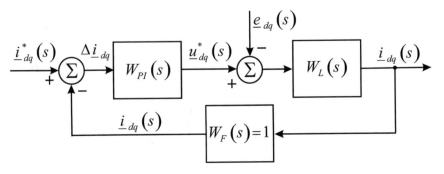

Fig. 4.6 Block diagram of the synchronous d-q frame PI controller

The transfer function of (4.21) has imaginary factors in denominator. Therefore, it can be expressed in terms of the real part W_{CLR} and the imaginary part W_{CLI}. In cases where the current reference has both the real part I_D^* and the imaginary part I_Q^*, the response of the d-axis current has one part proportional to the reference I_D^*, but also an additional part which is proportional to the reference I_Q^*. The same conclusion holds for the response of the q-axis current (4.23). The consequences of such undesired coupling are negative. In grid-side inverters, the step of the active power will also include an inadvertent transient of the reactive power and the other way around. In ac drives, the step of the electromagnetic torque will provoke an undesired transient of the flux and vice versa. The coupling depends on the factor $\omega_e L$, which resides in the denominator of (4.21). Thus, the coupling is larger at higher excitation frequencies;

$$
\begin{aligned}
i_d(s) &= W_{CLR}(s) \cdot I_D^*(s) - W_{CLI}(s) \cdot I_Q^*(s), \\
i_q(s) &= W_{CLI}(s) \cdot I_D^*(s) + W_{CLR}(s) \cdot I_Q^*(s).
\end{aligned}
\tag{4.23}
$$

4.3   Decoupling Current Controller in *d-q* Frame

In grid-side inverters, the current component of the *d-q* system corresponds to active and reactive power. In ac drives, they correspond to the flux and torque. In both cases, it is not desirable to have any coupling between the *d*-axis and the *q*-axis. With synchronous PI controller in *d-q* frame, the transients in one axis produce considerable effects in the other axis. In order to prevent the coupling, it is necessary to design the decoupling controller. The principles of the internal model control (IMC) are introduced and explained in *s*-domain.

4.3.1   Basic Principles of Internal Model Control

The goal of all the controllers is to make the output (i in Fig. 4.7) track the reference i^* with no errors. For this to achieve, the closed-loop transfer function should be equal to 1. Due to low-pass nature of the load W_L, it is not possible to obtain an instantaneous tracking of all the changes of the reference, in particular the step changes. The change of the output i is largely affected by the dynamics of the load W_L. Within the closed-loop transfer function, such as the one in (4.21), the load transfer function W_L gets multiplied by the transfer function of the controller. This provides the possibility to cancel out the undesired dynamics of the object by including W_L^{-1}, or at least some of its elements into the transfer function of the controller.

The load transfer function in d-q frame is given in (4.17). Inverted dynamics of the load are given in (4.24). With one zero and no poles in W_L^{-1}, an attempted design of the controller where $W_{REG} = W_L^{-1}$ has to be corrected by multiplying the inverted dynamics by α/s, where α is an adjustable parameter (4.25).

$$W_L^{-1}(s) = R + L \cdot s + j \cdot L \cdot \omega_e. \tag{4.24}$$

$$W_{REG}(s) = W_L^{-1}(s)\frac{\alpha}{s} = \frac{\alpha R}{s} + \alpha L + j\frac{\alpha L \omega_e}{s}. \tag{4.25}$$

4.3.2   Decoupling Controller

Proposed current controller of (4.25) resembles the PI controller. The proportional gain is αL, while the integral gain is αR. In addition, the third factor is an integrator with the gain $j\alpha L\omega_e$ proportional to the output frequency. The third gain is multiplied by the imaginary unit. This means that the error Δi_d will affect not only the voltage u_d^* but also the voltage u_q^*. At the same time, the error Δi_q will affect not only the voltage u_q^* but also the voltage u_d^*. Rewriting the current controller in its scalar form, where the real and the imaginary parts are separated, one obtains

$$
\begin{aligned}
u_d^* &= \left(\frac{\alpha R}{s} + \alpha L\right) \cdot \Delta i_d - j\frac{\alpha L\omega_e}{s} \cdot \Delta i_q, \\
u_q^* &= \left(\frac{\alpha R}{s} + \alpha L\right) \cdot \Delta i_q + j\frac{\alpha L\omega_e}{s} \cdot \Delta i_d.
\end{aligned}
\tag{4.26}
$$

With the current controller of (4.26), the closed-loop transfer function of the system in Fig. 4.7 is

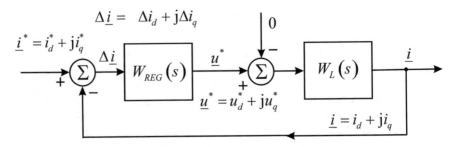

Fig. 4.7 In application of internal model control concept, the controller W_{REG} comprises the inverted dynamics of the object W_L

$$W_{CL}(s) = \frac{W_L(s) \cdot W_{\mathrm{REG}}(s)}{1 + W_L(s) \cdot W_{\mathrm{REG}}(s)} = \frac{\alpha}{s + \alpha}. \qquad (4.27)$$

The closed-loop transfer function does not have any imaginary parts. Thus, there is no coupling between the orthogonal axes of the d-q coordinate frame. At the same time, W_{CL} represents the first-order transfer function with the time constant $1/\alpha$, where the output approaches the reference input in an exponential manner, with the steady-state value which is equal to the reference. The controller (4.26) is also called the decoupling controller, as it prevents the undesired coupling between the d and q transients.

Practical implementation of the decoupling controller deals with the pulse transfer functions in z-domain, and it must take into account all the delays within the feedback acquisition path and the delays introduced by the PWM inverter. This implementation is discussed in the subsequent chapters. Due to delays, the z-domain pulse transfer function of the load $W_L(z)$ comprises the elements that cannot be inverted. Formal inversion of the delay implies prediction, and these predictions cannot be implemented. In such cases, it is necessary to identify all the elements of $W_L(z)$ that are invertible and to use them for the controller design in the same manner as demonstrated in (4.25).

4.4 Resonant Current Controllers

The implementation of synchronous, d-q frame controllers requires numerically intensive coordinate transformations. Before the availability of fast digital signal controllers, this circumstance was a serious disadvantage of the d-q frame controllers. At the same time, while the synchronous d-q frame current controllers efficiently suppress the current errors of the direct-sequence line-frequency currents, they cannot suppress the errors caused by the inverse-sequence voltages, which appear in nonsymmetrical three-phase systems. Moreover, they cannot suppress the low-order line harmonics such as the fifth or seventh, which revolve at the speed considerably larger than the speed of the d-q frame.

The current controllers in stationary α–β frame do not use the Park transformation, and their numerical requirements are lower. The PI current controller of Sect. 4.2.1 comprises an integrator. The integral control action K_I/s suppresses the steady-state errors encountered while tracking the Heaviside step references, the references of the form $1/s$. According to (4.20), the stationary-frame PI controller cannot track the sinusoidal references without the phase and amplitude errors. In this section, discussion is focused on *resonant controllers*, the current controllers that reside in the steady state and which are capable to remove any error at the selected frequency ω, whether the line frequency or the frequency of selected lower-order line harmonic.

4.4.1 Transformation of the d-q Frame Controller in α–β Frame

The resonant controllers can be introduced by considering the α–β representation of the d-q frame PI controller, the latter being capable of suppressing the errors in tracking the direct-sequence line-frequency current references. For the d-q frame PI controller of Fig. 4.6, the transfer function $W_{PI}(s)$ is given in

$$W_{PI}(s) = \frac{\underline{u}_{dq}(s)}{\Delta \underline{i}_{dq}(s)} = K_P + \frac{K_I}{s}. \tag{4.28}$$

The complex image $\underline{u}_{dq}(s)$ can be expressed in terms of the $\underline{u}_{\alpha\beta}(s)$:

$$\underline{u}_{dq}(s) = \int_{0^-}^{+\infty} \underline{u}_{dq}(t) \cdot e^{-st} dt = \int_{0^-}^{+\infty} \underline{u}_{\alpha\beta}(t) \cdot e^{-j\theta_e} \cdot e^{-st} dt$$

$$= \int_{0^-}^{+\infty} \underline{u}_{\alpha\beta}(t) \cdot e^{-j\omega_e t} \cdot e^{-st} dt = \int_{0^-}^{+\infty} \underline{u}_{\alpha\beta}(t) \cdot e^{-(j\omega_e + s)t} dt \tag{4.29}$$

$$= \underline{u}_{\alpha\beta}(s + j\omega_e).$$

Similarly, the complex image $\Delta \underline{i}_{dq}(s)$ can be expressed in terms of $\Delta \underline{i}_{\alpha\beta}(s)$, $\Delta \underline{i}_{dq}(s) = \Delta \underline{i}_{\alpha\beta}(s + j\omega)$. Therefore, the transfer function of the PI controller in d-q frame $W_{PI}(s) = \underline{u}_{dq}(s)/\Delta \underline{i}_{dq}(s)$ corresponds to the ratio $\underline{u}_{\alpha\beta}(s + j\omega)/\Delta \underline{i}_{\alpha\beta}(s + j\omega)$ in the stationary α–β frame. Therefore, the d-q frame PI controller can be implemented in α–β frame by adopting the controller transfer function

$$W_{REG1}^{\alpha\beta}(s) = K_P + \frac{K_I}{s - j\omega_e}. \tag{4.30}$$

Hence, instead of using the PI controller in d-q frame, where the controller receives $\Delta i_{dq}(s)$ and provides $u_{dq}(s)$, it is possible to implement the current controller in α–β frame, where it receives $\Delta i_{\alpha\beta}(s)$ and provides $u_{\alpha\beta}(s)$. As long as the transfer function of the current controller in α–β frame corresponds to (4.30), the system behaves in the same way as the system with d-q frame PI controller. Among other properties, such controller is capable of controlling the direct-sequence ac currents at the line frequency without any amplitude or phase errors.

In parallel with the PI current controller implemented in d-q frame which revolves at the speed ω_e in positive direction, it is possible to implement the PI controller in d_i-q_i frame, the one that revolves at the same speed in negative direction. If the speed ω_e corresponds to the line frequency, then the PI controller in d_i-q_i frame manages to suppress any current error in the inverse sequence of line-frequency ac currents. Relying on the results (4.29) and (4.30), the PI controller in d_i-q_i frame can be replaced by the current controller in the stationary α–β frame, provided that the transfer function of such controller is

$$W_{REG2}^{\alpha\beta}(s) = K_P + \frac{K_I}{s + j\omega_e}. \tag{4.31}$$

Parallel operation of current controllers (4.30) and (4.31) can be achieved by adding the voltage references obtained at their outputs. The equivalent transfer function of such double-track current controller is

$$\begin{aligned} W_{REG12}^{\alpha\beta}(s) &= \frac{u_{\alpha\beta}(s)}{\Delta i_{\alpha\beta}(s)} = W_{REG1}^{\alpha\beta}(s) + W_{REG2}^{\alpha\beta}(s) \\ &= K_P + \frac{K_I}{s - j\omega_e} + K_P + \frac{K_I}{s + j\omega_e} \\ &= 2 \cdot K_P + \frac{2 \cdot s \cdot K_I}{s^2 + \omega_e^2}. \end{aligned} \tag{4.32}$$

The current controller (4.32) is implemented in α–β frame, and it manages to suppress any current error when operating with ac current of the frequency ω_e, including both direct and inverse sequences.

4.4.2 The Resonant Controller in α–β Frame

In (4.32), the structure of the controller resembles the PI controller. The factor $1/s$ of the integrator is replaced by the factor $s/(s^2 + \omega^2)$. In this way, the capability of removing the errors created by the Heaviside input-step $(1/s)$, the controller is capable of removing the errors while tracking the ac currents of the frequency ω. The new integrator comprises the resonant tank; therefore, the controller is also called the *resonant controller*.

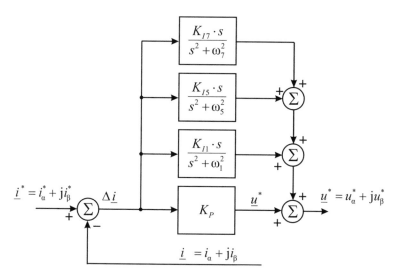

Fig. 4.8 The resonant controller tuned for the fundamental frequency and for the fifth and seventh harmonic

Considering the transfer function of the load in α–β frame, $W_L(s) = 1/(R + sL)$, and the transfer function (4.32) of the resonant controller, the closed-loop transfer function in α–β frame is

$$W_{CL}(s) = \frac{W_L(s) \cdot W_{REG12}^{\alpha\beta}(s)}{1 + W_L(s) \cdot W_{REG12}^{\alpha\beta}(s)}$$

$$= \frac{2K_P(s^2 + \omega_e^2) + 2K_I s}{(R + sL) \cdot (s^2 + \omega_e^2) + 2K_P(s^2 + \omega_e^2) + 2K_I s}. \tag{4.33}$$

For the excitation frequency $s = \pm j\omega_e$, the closed-loop transfer function becomes $W_{CL}(\pm j\omega_e) = 1$. Thus, there is no steady-state error in controlling the currents of the frequency ω_e.

One resonant controller can use several factors of the form $s/(s^2 + \omega^2)$, each having a different frequency ω, contributing to elimination of several line harmonics. An example of such resonant controller is given in Fig. 4.8.

The presence of resonant tanks within the controller transfer function affects the step response of the closed-loop current controller. In the next section, transient properties of resonant controllers are studied and explored using a simplified model in s-domain.

4.4.3   Dynamic Properties of the Resonant Controller

The presence of resonant integrators within the current controller of Fig. 4.8 suppresses the direct and inverse components of the current error, as well as selected line

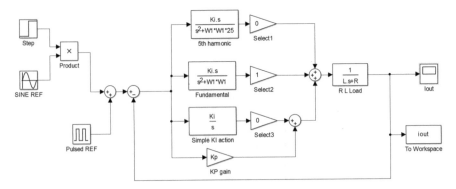

Fig. 4.9 Simulation of the resonant current controller with resonant integrators adjusted for the fundamental and for the fifth harmonic

harmonics. At the same time, the resonant integrators alter the step response. In order to explore the impact of the resonant elements on the step response of the current controller, dynamic response of the resonant controller is simulated on the computer.

In Fig. 4.9, the simulation model comprises the current reference generator; the current controller with proportional, integral, and resonant actions; and the model of the load. By means of the selection switches, the model can be configured to operate as:

- The PI current controller with the closed-loop bandwidth of 250 Hz
- The resonant controller tuned for the line-frequency currents
- The resonant converter which also suppresses the fifth harmonic

The input step response is tested in all three configurations. The simulation traces are given in Fig. 4.10. The upper trace represents the step response of the PI controller in the stationary α–β frame, tuned to achieve the closed-loop bandwidth of 250 Hz. The trace in the middle corresponds to the resonant controller which suppresses the current errors at the fundamental, line frequency. The lower trace is obtained with the resonant current controller which also includes the resonant integrator tuned for the fifth line-frequency harmonic.

The traces in Fig. 4.10 present the step responses that comprise weakly damped oscillations at both resonant frequencies, the fundamental line frequency and the frequency of the fifth harmonic. The oscillations are caused by the resonant integrators which get excited by the step input. Although the controller provides suppression of the ac errors in steady-state conditions, stepwise changes of the input reference give rise to oscillations which deteriorate the step response of the system.

There are applications where the use of resonant controllers is justifiable, notwithstanding the parasitic oscillations. In other applications, it is of interest to find alternative ways to suppress the errors caused by the inverse component within the line voltages and to reduce the effects of the line-frequency harmonics.

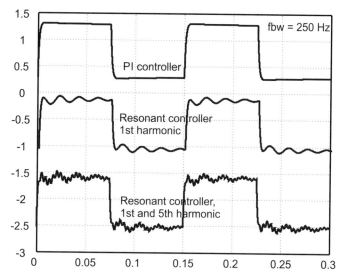

Fig. 4.10 Step response of the PI controller, resonant current controller tuned for the line-frequency fundamental component, and resonant controller tuned for the fundamental and the fifth harmonic

4.5 Disturbance Rejection

In addition to changes caused by the input reference, the output current also changes in response to the voltage disturbances, such as the variation in the back-electromotive force or the changes in line voltages. The current changes caused by the voltage disturbances should be as low as possible, preferably zero. The capability to suppress the disturbances is characterized by the output admittance of the current controller, $Y(j\omega) = \Delta i(j\omega)/\Delta u(j\omega)$, where Δi is the current error while Δu is the voltage disturbance. It is desirable to obtain the values $Y(j\omega)$ as low as possible over the whole frequency range where the voltage disturbances could take place. The value of $Y(j\omega)$ can be reduced by using additional control action called the *active resistance feedback*, discussed in this section.

4.5.1 Disturbance Rejection with d-q Frame PI Controller

The block diagram of the decoupling current controller in *d-q* frame is given in Fig. 4.11, along with the transfer function of the load and the transfer function $W_{REG}(s)$. The impact of the load disturbance on the output current is obtained by considering the ratio $i_{dq}(s)/e_{dq}(s)$ in cases where $i^*_{dq}(s) = 0$. From Fig. 4.11, desired ratio is given in (4.34).

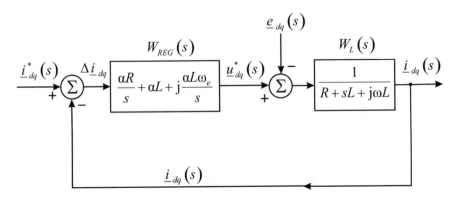

Fig. 4.11 Decoupled PI controller in d-q coordinate frame

$$Y_{dq}(s) = \frac{i_{dq}(s)}{e_{dq}(s)}\bigg|_{i^*_{dq}=0} = \frac{-W_L(s)}{1 + W_L(s)W_{REG}(s)} = \frac{-s}{(s+\alpha)\cdot(R+sL+jL\omega_e)}.$$

(4.34)

According to considerations outlined in Sect. 4.4.1, the transfer function $W_{\alpha\beta}(s)$ from the stationary frame corresponds to the transfer function $W_{dq}(s+j\omega)$ in the synchronous d-q frame, while $W_{dq}(s)$ corresponds to $W_{\alpha\beta}(s-j\omega)$. Therefore, the output admittance $Y_{\alpha\beta}(s)$ in the stationary frame becomes

$$Y_{\alpha\beta}(s) = Y_{dq}(s - j\omega_e) = \frac{-s + j\omega_e}{(s - j\omega_e + \alpha)\cdot(R + sL)}.$$

(4.35)

Notice that the value of $Y_{\alpha\beta}(s)$ is equal to zero for the direct-sequence voltage disturbances, where $s = j\omega_e$. Thus, the direct-sequence, line-frequency voltage disturbances do not cause any steady-state errors in the output current. Namely, the system is capable of suppressing the impact of these disturbances completely. The amplitude characteristic of $Y_{\alpha\beta}(j\omega)$ is given in Fig. 4.12, where the frequency sweeps from the negative-sequence eleventh harmonic up to the positive-sequence eleventh harmonic. The plot is obtained for the decoupling d-q frame controller set for the closed-loop bandwidth of 250 Hz and with $L = 10$ mH and $R = 5\ \Omega$. For convenience, the plotted values correspond to the logarithm of the admittance $Y_{\alpha\beta}(j\omega)$. At the fundamental frequency $s = j\omega_e = j\cdot314.15$ rad/s, the amplitude of $Y_{\alpha\beta}(j\omega)$ drops to zero. In the region of the fifth harmonic, the value of $|Y_{\alpha\beta}(j\omega)|$ is close to 0.037. This means that the voltage disturbance of 100 V at the frequency of the fifth harmonic will give a rise to the erroneous fifth harmonic current of 3.7 A. It is of interest to provide the means to reduce the values of $|Y_{\alpha\beta}(j\omega)|$ and thus reduce the current errors caused by the voltage disturbances.

A numerical indicator of the disturbance rejection capability is the ability of the current controller to reject the zero-frequency components in α–β frame. In Fig. 4.12, the corresponding point is denoted by A. The dc disturbance in stationary

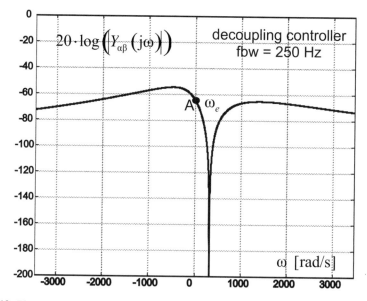

Fig. 4.12 The amplitude characteristic of the output admittance in α–β frame

α–β frame can be transformed in synchronous, d-q frame, where it becomes an ac disturbance which revolves at the speed of $-j\omega_e$. The capability to suppress this disturbance is determined by $Y_{dq}(-j\omega_e)$, which is given in (4.36):

$$Y_{dq}(-j\omega_e) = \frac{j\omega_e}{(-j\omega_e + \alpha) \cdot R}, \qquad |Y_{dq}(-j\omega_e)| \approx \frac{\omega_e}{\alpha R}. \qquad (4.36)$$

The closed-loop bandwidth of the system is determined by α, which is considerably larger than the line-frequency ω_e. Therefore, the admittance in (4.36) reduces to $\omega_e/(\alpha R)$. Parameter R corresponds to the equivalent series resistance of the inverter, inductances, and the grid. The value of R is usually rather low, in particular with large power grid-side converters. For this reason, the output admittance can be considerable, leading to a low disturbance rejection capability.

4.5.2 Active Resistance Feedback

The admittance $\omega_e/(\alpha R)$ of (4.36) is inversely proportional to the equivalent series resistance. A larger value of R results in a more efficient suppression of the voltage disturbances. An attempt to increase R would also produce the undesirable power losses. Instead of increasing the actual series resistance, it is possible to introduce a new control action that would have the same impact on the disturbance suppression capability, yet without the need to increase an actual resistance that increases the losses.

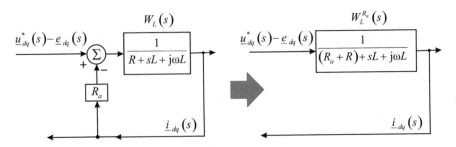

Fig. 4.13 The implementation of the active resistance feedback

The introduction of a local proportional control action can increase the parameter R within the transfer function W_L. The left side schematic in Fig. 4.13 comprises the local feedback with the proportional gain R_a. It consists in changing the voltage command by subtracting the factor $R_a \cdot i_{dq}(s)$.

The equivalent transfer function of the load is shown in the right side schematic in Fig. 4.13. The equivalent resistance of the load subsystem is increased by R_a, and it becomes $R + R_a$. At the same time, the admittance of (4.36) becomes $\omega_e/(\alpha R + \alpha R_a)$. With sufficiently large R_a, disturbance rejection capability can be significantly increased. In practical implementation, the range of applicable gains R_a is limited by the feedback acquisition delays, the PWM actuator delays, and the noise and unmodeled dynamics. These aspects and practical R_a limits are discussed in subsequent chapters.

Question (4.1) The load inductance is equal to $L = 10$ mH, while the winding resistance is $R = 10\ \Omega$. The current is controlled by the PI current controller located in the stationary coordinate frame (α–β). The closed-loop transfer function is given in (4.18). It is necessary to find the gains K_P and K_I that produce two real poles of (4.18) at $f_p = 500$ Hz. The current controller is used to control the line-frequency currents injected into 50 Hz grid. Determine the amplitude and phase error obtained in the steady state.

Answer (4.1) In order to obtain two real poles at $\omega_p = 2\pi\ 500$ rad/s, the coefficients of denominator of (4.18) have to be $L/K_I = 1/(\omega_p)^2$ and $(R + K_P)/K_I = 2/\omega_p$. From there, $K_I = 98{,}700$, $(R + K_P) = 62.83$, and $K_P = 52.83$. From (4.20), the amplitude characteristic deviates by 0.4%, while the phase lag is 1.87 degrees.

Question (4.2) The load inductance is equal to $L = 10$ mH, while the winding resistance is $R = 5\ \Omega$. The current is controlled by the decoupling current controller located in synchronous coordinate frame. The current controller is used to control the line-frequency currents injected into 50 Hz grid. It is necessary to achieve the closed-loop bandwidth of 500 Hz. It is necessary to determine the control actions of the decoupling current controller and to calculate the closed-loop transfer function.

Answer (4.2) The closed-loop transfer function of the decoupling current controller is given in (4.27), while the transfer function of the controller is given in (4.26).

In order to achieve the bandwidth of 500 Hz, the gain should be $\alpha = 2\pi\ 500$. The proportional gain is $\alpha L = 10\ \pi$, while the integral gain is $\alpha R = 5000\ \pi$. The gain of the decoupling factor $j\alpha L\omega_e$ is equal to $10\ \pi\ 2\ \pi\ 50 = 9870$.

Question (4.3) Decoupling current controller is implemented in d-q coordinate frame. The load inductance is equal to $L = 10$ mH, while the winding resistance is $R = 5\ \Omega$. The current controller is used to control the line-frequency currents injected into 50 Hz grid. The controller achieves the closed-loop bandwidth of 400 Hz. The line voltage comprises the seventh harmonic with an amplitude of 50 V. It is necessary to determine the current error caused by the seventh harmonic voltage disturbance.

Answer (4.3) The output impedance in α-β frame is given in (4.35). In order to achieve the bandwidth of 400 Hz, the gain should be $\alpha = 2\pi\ 400$. The speed of the d-q system is $\omega_e = 2\pi\ 50$. By introducing $s = j\ 7\ \omega_e$ in (4.35), one obtains $Y_{\alpha\beta7} = -0.0243 + j\ 0.0108$. The absolute value of the admittance for the seventh harmonic is 0.0266. Therefore, the current will comprise the seventh harmonic of 1.33 A. These considerations take into account the voltage disturbance that has the direct sequence of the seventh harmonic. In cases where the seventh harmonic voltages revolve in the opposite direction, the value of $Y_{\alpha\beta7}$ becomes $Y_{\alpha\beta7} = -0.0265 + j\ 0.0167$, the absolute value is 0.0314, while the seventh harmonic of the output current is 1.57 A.

Question (4.4) Consider the three-phase R-L load with symmetrical back-electromotive force. The α-β frame current controller has K_P and K_I control actions. The voltage actuator can be replaced by the unity gain (the output voltage is equal to the reference). The back-electromotive force is the voltage disturbance with adjustable angular frequency ω_D. The amplitude of the back-electromotive force is E_D. It is necessary to determine the current error I_D caused by the voltage disturbance E_D.

Answer (4.4) The block diagram of the system with stationary-frame PI controller is given in Fig. 4.4. The transfer function of the load is $W_L(s) = 1/(R + L\ s)$. The transfer function of the PI current controller is $W_{PI}(s) = K_P + K_I/s$. In cases where the current reference is equal to zero, the output current $i(s)$ depends on the voltage disturbance $E_D(s)$:

$$-W_{PI}(s) \cdot i(s) - E_D(s) = \frac{i(s)}{W_L(s)},$$

$$i(s) = -\frac{W_L(s)}{1 + W_{PI}(s) \cdot W_L(s)} E_D(s),$$

$$i(s) = -\frac{\dfrac{1}{R + sL}}{1 + \dfrac{K_I + sK_P}{s} \cdot \dfrac{1}{R + sL}} E_D(s),$$

$$i(s) = -\frac{s}{s^2 L + s(R + K_P) + K_I} E_D(s),$$

$$i(j\omega_D) = -\frac{j\omega_D}{(K_I - L\omega_D^2) + j\omega_D(R + K_P)} E_D(j\omega_D).$$

Question (4.5) Desired closed-loop transfer function for the current controlled system of Question 4.4 is $W_{SS}(s) = 1/(1 + s/\alpha)$, where $\alpha = 1000$ rad/s. The load parameters are $L = 10$ mH and $R = 10\ \Omega$. Calculate the current controller gains K_P and K_I.

Answer (4.5) In order to achieve the closed-loop transfer function $W_{SS}(s) = 1/(1 + s/\alpha)$, the open-loop gain $W_{PI}(s)W_L(s)$ has to be equal to α/s. Therefore, the transfer function of the PI controller has to be $W_{PI}(s) = \alpha R/s + \alpha L$, while $K_P = \alpha L = 10$ and $K_I = \alpha R = 10000$.

Question (4.6) For the system of Questions 4.4 and 4.5, the voltage disturbance E_D is 100 V and the frequency of the disturbance is $\omega_D = 100$ rad/s. Calculate the amplitude of the current error, caused by such a disturbance.

Answer (4.6) According to developments in Question 4.4 and with the parameters of Question 4.5,

$$
\begin{aligned}
i(j\omega_D) &= -\frac{j\omega_D}{(K_I - L\omega_D^2) + j\omega_D(R + K_P)} E_D(j\omega_D) \\
&= -\frac{j100}{(K_I - L \cdot 10000) + j100(R + K_P)} 100 \\
&= -\frac{j100}{(10000 - 100) + j100(10 + 10)} 100 \\
&= -\frac{j10000}{9900 + j2000}, \qquad |i(j\omega_D)| = 0.99\text{A}
\end{aligned}
$$

Question (4.7) The system of Question 4.4 is enhanced by the active resistance feedback, suited to reduce the impact of the voltage disturbance on the output current. The value of the active resistance gain is set to R_a. It is necessary to set the gains of the PI controller to achieve the closed-loop transfer function $W_{SS}(s) = 1/(1 + s/\alpha)$.

Answer (4.7) The active resistance feedback is a local proportional gain that adds the value $-R_a i$ to the voltage command. Considering the load $W_L(s)$ with active resistance feedback, the equivalent transfer function of modified load is

$$
W_{ML}(s) = \frac{W_L(s)}{1 + W_L(s) \cdot R_a} = \frac{\dfrac{1}{R + sL}}{1 + \dfrac{R_a}{R + sL}} = \frac{1}{(R_a + R) + sL}.
$$

For the load transfer function $W_{ML}(s)$, it is necessary to set the gains of the PI controller so as to obtain $W_{ML}(z)W_{PI}(s) = \alpha/s$. Therefore, the transfer function of the PI controller has to be $W_{PI}(s) = \alpha(R + R_a)/s + \alpha L$. Therefore, $K_P = \alpha L$ and $K_I = \alpha(R + R_a)$.

Question (4.8) The system of Question 4.7 has the load parameters $L = 10$ mH and $R = 10\ \Omega$. The value of the active resistance gain is set to $R_a = 9R$. The value of α is equal to 1000 rad/s. The voltage disturbance E_D is 100 V and the frequency of the disturbance is $\omega_D = 100$ rad/s. It is necessary to determine the current error I_D caused by the voltage disturbance E_D.

Answer (4.8) From the results in Question 4.7, the gains are equal to $K_P = \alpha L = 10$ and $K_I = \alpha(R + R_a) = 100000$. According to developments in Question 4.4 and Question 4.6, the current error is equal to

$$
\begin{aligned}
i(j\omega_D) &= -\frac{j\omega_D}{(K_I - L\omega_D^2) + j\omega_D(R + R_a + K_P)} E_D(j\omega_D) \\
&= -\frac{j100}{(K_I - L \cdot 10000) + j100(R + R_a + K_P)} 100 \\
&= -\frac{j100}{(100000 - 100) + j100(10 + 90 + 10)} 100 \\
&= -\frac{j10000}{99900 + j11000}, \quad |i(j\omega_D)| = 0.099\text{A}
\end{aligned}
$$

Chapter 5
Discrete-Time Synchronous Frame Controller

This chapter provides the analysis, design, parameter setting, and evaluation of synchronous frame decoupling current controller. Discussion includes all the practical aspects of discrete-time implementation, and it includes discrete-time nature of the PWM voltage actuator and the feedback acquisition systems which are described in the preceding chapters. The analysis and design of discrete-time current controllers rely on z-transform, and it takes into account all the transport delays caused by the feedback acquisition, computation, and PWM processes. Although the scope does not include the code writing, some basic notions related to the interrupt execution and scheduling are also taken into account.

Design and parameter setting of the discrete-time controller are performed for two distinct cases, which have different methods of acquiring the feedback signals. The first is the center-pulse sampling technique, wherein the PWM pulses are symmetrical and the feedback samples are captured in the middle of the voltage pulses. This technique is simple to implement, but the use of only one sample makes the feedback prone to the PWM noise and parasitic phenomena. The second technique is oversampling-based, and it derives the feedback from the train of samples acquired within the past PWM period. This one-PWM-period averaging improves the robustness of the controller against the noise in the feedback path. Any PWM-related noise is removed, improving at the same time the robustness against spurious and parasitic noise sources.

For both feedback acquisition techniques, the structure of the discrete-time current controller is designed using the IMC approach. The parameter setting is suited to maximize the closed-loop bandwidth, yet preserving an overshoot-free response and sufficient robustness against the parameter changes. An insight into the closed-loop performance of both controllers is obtained analytically, by means of computer simulations and also from the experimental results.

With both feedback acquisition schemes, the transport delays are the limiting factor of the closed-loop performance. The effect of delays can be reduced by introducing the series compensator with differential action. Both current controllers discussed in this chapter are extended in order to reduce the effects of the transport

© Springer International Publishing AG, part of Springer Nature 2018
S. N. Vukosavic, *Grid-Side Converters Control and Design*, Power Electronics and Power Systems, https://doi.org/10.1007/978-3-319-73278-7_5

delays. The closed-loop performance of conventional and extended current controllers is evaluated by means of computer simulations and by experimental runs.

5.1 Discrete-Time Controller with Center-Pulse Sampling

Characteristics of the discrete-time current controller largely depend on the method which is used to acquire the feedback signals. The center-pulse sampling technique operates in conjunction with symmetrical PWM, where the centers of all the voltage pulses in all the phases coincide. It has been shown in the previous chapters that the PWM-related current ripple crosses zero at the center of the voltage pulses. Therefore, the sampling is performed at the center of the voltage pulses. This technique is simple to implement, and it does not require excessive computations. Yet, the use of only one sample makes the feedback prone to the PWM noise and parasitic phenomena. The first step of the design is the calculation of the discrete-time model of the load. Based on the pulse transfer function of the load, the current controller structure is designed using the IMC approach.

5.1.1 The Pulse Transfer Function of the Load

Analysis and design of current controllers are simplified by representing the output current as a vector. The vector of the output current can be expressed in terms of the unit vectors in the stationary $\alpha-\beta$ frame, as well as unit vectors in the synchronous d-q frame. For the convenience, the notation can be changed; instead of using the unit vectors, the output current in $\alpha-\beta$ frame can be written as a complex number, where the real part is the projection of the vector in α-axis, while the imaginary component is the projection on β-axis. For brevity, in all further considerations, the output current in $\alpha-\beta$ frame is denoted by $i^s = i_{\alpha s} + ji_{\beta s}$. Similarly, the output current in the synchronous d-q frame is denoted by $i^e = i_d + ji_q$. The superscript s is also used for other variables in the stationary $\alpha-\beta$ frame, while the superscript e is used for the variables in the synchronous d-q frame.

In Fig. 5.1, the sampling period T_S is equal to one half of the PWM period. The feedback samples i_n are taken at the zero-count and the period-count of the triangular PWM carrier, at instants that coincide with the center of the voltage pulses. At each such instant nT_S, an interrupt event starts the execution of the control routine, denoted by *EXE* in Fig. 5.1. The interrupt routine triggered at nT_S takes the feedback i_n, performs the necessary transformations, discriminates the current error, executes the relations of the controller, and calculates the voltage reference u_n^*. The voltage reference that is calculated in interrupt routine triggered at nT_S cannot be applied within the same sampling period. At the time when the *EXE* routine ends, and the voltage reference becomes available, the desired commutation instant may have already passed.

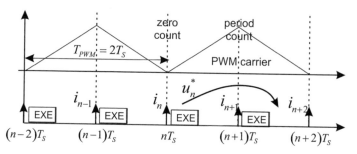

Fig. 5.1 Scheduling of the current controller tasks

Due to the PWM delay, the reference u_n^* gets applied within the successive interval $[(n + 1)T_S \,.. \, (n + 2)T_S]$. Similarly, the reference u_{n-1}^* gets applied within the interval $[nT_S \,.. \, (n + 1)T_S]$.

The differential equation that describes the change of the current of the system in Fig. 4.1 is

$$Ri^s(t) + L\frac{di^s(t)}{dt} = u^s(t) - e^s(t),\qquad(5.1)$$

where $e^s(t)$ is the back-electromotive force in α–β frame, R is the load resistance, and L is the load inductance. Assuming that the $e^s{}_n$ stands for the average value of the electromotive force over the same interval, and introducing $\beta = RT_S/L$, the difference equation expressing the change of the stator current within the interval $[nT_S \,.. \, (n + 1)T_S]$ becomes

$$i_{n+1}^s = i_n^s e^{-\beta} + \frac{1 - e^{-\beta}}{R}\left(u_{n-1}^s - e_n^s\right)\qquad(5.2)$$

In order to obtain the model of the load in d-q coordinate frame, it is necessary to transform the relation (5.2) from α–β into the d-q frame. Complex vectors in stationary and synchronous frames are related by

$$i_{n+1}^s = i_{n+1}^e \cdot e^{j\theta_{n+1}},\, i_n^s = i_n^e \cdot e^{j\theta_n},\, u_{n-1}^s = u_{n-1}^e \cdot e^{j\theta_{n-1}},\qquad(5.3)$$

where θ_n represents the angle between the d-axis and the α-axis at instant nT_S. Assuming that the speed of the d-q frame is $\omega_e = d\theta/dt$ and that the speed changes within the interval T_S are negligible, the changes of the d-q frame angle within one sampling instant are expressed by $\theta_{n + 1} = \theta_n + \omega_e T_S$ and $\theta_n = \theta_{n-1} + \omega_e T_S$. The average value of the electromotive force within the interval $[nT_S \,.. \, (n + 1)T_S]$ becomes

$$e_n^e = \frac{1}{T}\int_{nT_S}^{(n+1)T_S} e^e(t) \cdot dt \approx e_n^s \cdot e^{-j(\theta_{n+1}+\theta_n)/2}.\qquad(5.4)$$

By introducing the results (5.3) and (5.4) into the difference Eq. (5.2) and dividing the result by $\exp(j\theta_n)$, one obtains

$$i_{n+1}^e \cdot e^{j\omega_e T_S} = i_n^e \cdot e^{-\beta} + \frac{1-e^{-\beta}}{R}\left(u_{n-1}^e \cdot e^{-j\omega_e T_S} - e_n^e \cdot e^{j\omega_e T_S/2}\right). \tag{5.5}$$

The electrical time constant L/R of the load is often larger than 5–10 ms, while the sampling time with $f_{PWM} = 10$ kHz is $T_S = 50$ µs. Therefore, the factor $\beta = RT_S/L$ is lower than 1/100. Therefore, the value $(1-e^{-\beta})/R$ can be approximated T_S/L, and the expression (5.5) becomes

$$i_{n+1}^e \cdot e^{j\omega_e T_S} = i_n^e \cdot e^{-\beta} + \frac{T_S}{L}\left(u_{n-1}^e \cdot e^{-j\omega_e T_S} - e_n^e \cdot e^{j\omega_e T_S/2}\right). \tag{5.6}$$

The difference Eq. (5.6) can be transformed in an algebraic equation in z-domain, where the variables are the complex images $i^e(z)$, $u^e(z)$, and $e^e(z)$. The operator z denotes a time advance of one sampling period. Thus, i_{n+1}^e transforms into $zi^e \cdot (z)$. With that in mind,

$$z \cdot i^e(z) \cdot e^{j\omega_e T_S} = i^e(z) \cdot e^{-\beta} + \frac{T_S}{L}\left(\frac{u^e(z)}{z \cdot e^{j\omega_e T_S}} - e^e(z) \cdot e^{j\omega_e T_S/2}\right). \tag{5.7}$$

The load transfer function in z-domain is obtained by dividing the complex images $i^e(z)$ and $u^e(z)$ in conditions where the disturbance $e^e(z)$ is equal to zero. From (5.7),

$$W_L(z) = \frac{i^e(z)}{u^e(z)}\bigg|_{e^e=0} = \frac{T_S}{L}\frac{1}{z \cdot e^{j\omega_e T_S}\left(z \cdot e^{j\omega_e T_S} - e^{-\beta}\right)}. \tag{5.8}$$

When both the voltage $u^e(z)$ and the back-electromotive force $e^e(z)$ are taken into consideration, the complex image of the output current in d-q frame is

$$i^e(z) = W_L(z) \cdot u^e(z) - z \cdot e^{j\omega_e T_S \cdot 3/2} W_L(z) \cdot e^e(z). \tag{5.9}$$

The expression (5.9) can be used in designing the structure of the current controller in d-q frame.

5.1.2 Design of the Controller Structure

The block diagram of the current controller implemented in d-q frame, which uses the synchronous, center-pulse sampling, is given in Fig. 5.2. With the feedback samples taken at the center of the voltage pulses and assuming that the delay caused by the anti-aliasing filter is negligible, the train of i^e pulses can be used as the train of i^{FB} pulses, $i^{FB}(z) = i^e(z)$. Thus, the feedback path has the gain of 1, and it has no

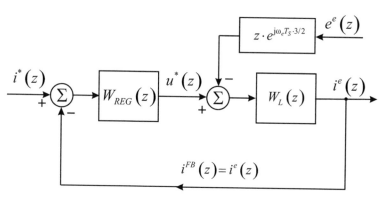

Fig. 5.2 The block diagram of the d-q frame current controller which uses the synchronous sampling scheme

delays. The PWM delays are taken into account in developing the pulse transfer function of the load (5.9). In Fig. 5.2, the input disturbance is the current reference i^* $(z) = i^*_d(z) + ji^*_q(z)$, the voltage disturbance is $e^e(z) = e^e_d(z) + je^e_q(z)$, while the output is $i^e(z) = i^e_d(z) + ji^e_q(z)$.

As discussed in the previous chapter, the IMC design consists in inverting dynamics of $W_L(z)$ and multiplying the result by the transfer function of an integrator. In z-domain, the pulse transfer function of an integrator is $1/(1-z^{-1}) = z/(z-1)$. The inverted dynamics of the load gets multiplied by $\alpha_D \cdot z/(z-1)$, where α_D is an adjustable gain. An attempt to invert the pulse transfer function of the load and to multiply the outcome by the factor $\alpha_D \cdot z/(z-1)$ results in

$$W_{REGX}(z) = \frac{\alpha_D z}{z-1} \cdot W_L^{-1} = \frac{\alpha_D L}{T_S} \frac{z^2 \cdot e^{j\omega_e T_S} \left(z \cdot e^{j\omega_e T_S} - e^{-\beta}\right)}{z-1} \qquad (5.10)$$

The expression (5.10) includes the factor $\exp(j\omega_s T_S)$, which makes the proper understanding of the proposed controller more difficult. For the purpose of discussion, it is assumed that $\omega_s = 0$ and that the factor z^2 can be disregarded. This reduces (5.10) to the form

$$W_{REGX}(z) = \left(\frac{\alpha_D L}{T_S}\right) \frac{z - e^{-\beta}}{z-1} = \frac{\alpha_D L}{T_S} + \frac{\alpha_D L}{T_S} \frac{1 - e^{-\beta}}{z-1}. \qquad (5.11)$$

In (5.11), the first factor $(\alpha_D L/T_S)$ is the proportional gain, while the second factor represents a delayed integrator. Thus, the basic control actions of the proposed controller (5.10) resemble the PI controller. It is of interest to recall the conclusions (4.26) of the s-domain implementation of the IMC concept. The proportional gain in (4.26) gets equal to α_L, where α is desired closed-loop bandwidth of the s-domain current controller in [rad/s]. By comparing the expressions (4.26) and (5.11), the gain α_D of the discrete-time controller corresponds to αT_S. The first estimate of the gain α_D, required to achieve the closed-loop bandwidth of $a = 2 \cdot \pi \cdot 1000 = 6283$ rad/sin the system with $f_{PWM} = 10$ kHz and $T_S = 50$ μs, is $\alpha_D = 0.31$.

Notice in (5.10) that the pulse transfer function $W_{REGX}(z)$ cannot be implemented, since it implies prediction. Namely, the order of the polynomial in numerator is 3, while the order of the polynomial in denominator is 1. In order to obtain the practical current controller, the pulse transfer function $W_{REGX}(z)$ has to be divided by z^2,

$$W_{REG}(z) \;=\; W_{REGX}(z)\frac{1}{z^2}.$$

Thus, the practical implementation becomes

$$W_{REG}(z) = \frac{\alpha_D L}{T_S}\frac{e^{j\omega_e T_S}\left(z \cdot e^{j\omega_e T_S} - e^{-\beta}\right)}{z - 1} \tag{5.12}$$

Notice in (5.13) that the proposed transfer function of the current controller has both the real and imaginary parts. In turn, the current error in d-axis would affect both the d-voltage command and the q-voltage command. With (5.13), the open-loop gain becomes

$$W_{OLG}(z) = W_{REG}(z)W_L(z) = \frac{\alpha_D}{z(z-1)}. \tag{5.13}$$

The closed-loop transfer function of the IMC-based controller is given in (5.14). The characteristic polynomial $f(z) = z^2 - z + \alpha_D$ has two poles and only one gain:

$$W_{CL}(z) = \frac{i^e(z)}{i^*(z)}\bigg|_{e^e=0} = \frac{W_{OLG}(z)}{1 + W_{OLG}(z)} = \frac{\alpha_D}{z^2 - z + \alpha_D}. \tag{5.14}$$

5.1.3   Parameter Setting

For $\alpha_D < 0.25$, the characteristic polynomial $f(z) = z^2 - z + \alpha_D$ has two real roots, z_1 and z_2, both of them positive, real numbers within the interval [0 1]. These roots are the poles of the closed-loop transfer function. The poles in z-domain and the corresponding poles in s-domain are correlated by $z_1 = \exp(s_1 T_S)$ and $z_2 = \exp(s_2 T_S)$. Thus, the abovementioned real poles correspond to the real s-domain poles that reside on the negative part of the real axis and extend between $-\infty$ and 0. Conjugate complex poles in z-domain map into conjugate complex poles in s-domain.

The values of α_D larger than 0.25 result in conjugate complex poles with the damping factor which decreases with the gain. The step response of the closed-loop system is plotted in Fig. 5.3 for the gain α_D ranging from 0.2 up to 0.4.

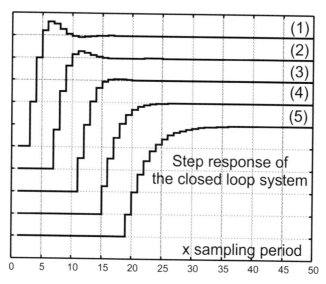

Fig. 5.3 Step response of the closed-loop system of (5.14) obtained with the gain a_D equal to 0.4 (trace 1), 0.35 (trace 2), 0.3 (trace 3), 0.25 (trace 4), and 0.2 (trace 5)

The traces (4) and (5) are obtained with $a_D = 0.25$ and $a_D = 0.2$. The closed-loop poles are real, and the step response does not have any overshoot. The trace (3) is obtained with $a_D = 0.30$. The closed-loop poles are conjugate complex numbers $z_{1/2} = 0.5 \pm j \cdot 0.2236$, the step response is well damped, and it has a very small, negligible overshoot. The traces (1) and (2) are obtained with $a_D = 0.40$ and $a_D = 0.35$, the damping is lower, and the step response has a significant overshoot. In order to avoid the overshoot, it is necessary to keep the gain a_D below 0.3.

The above discussion is obtained under assumption that the parameters of the load are constant. In practical application, the load parameters can change. One such parameter is the load inductance, which can reduce at larger current levels due to the magnetic saturation. Thus, the current controller should have the ability to provide stable responses even in cases where the load parameters change. The robustness of the current controller should be sufficient for the parameter changes of $\pm 20\%$ or even more.

The robustness of the closed-loop controller can be measured by the *vector margin* (VM). The vector margin can be explained by considering the closed-loop transfer function of (5.14). Instability is reached in cases where the closed-loop transfer function $W_{CL}(z)$ assumes an infinitely large value. These conditions are met if denominator $1 + W_{OLG}(z)$ becomes equal to zero. In order to avoid instability, the value of $W_{OLG}(z)$ must never be equal to -1. Thus, the stability margin can be judged from the lowest value of $|1 + W_{OLG}(z)|$. Assuming that the frequency ω sweeps from $-\infty$ to $+-\infty$ and taking into account the relation $z = \exp(j\omega T_S)$, the values of $W_{OLG}(z)$ are represented by the graph in the z-plane. The value of $|1 + W_{OLG}(z)|$ is lowest when the graph $W_{OLG}(\exp j\omega T_S))$ comes at the closest distance from the point $(-1, 0)$ of the z-plane. Such value is denoted by VM, and it is called the vector margin.

Table 5.1 Closed-loop performance of the decoupling controller

Gain α_D	VM	$f_{BW}(45°)$	$f_{BW}(-3\ dB)$	Overshoot
0.20	0.7608	$0.0263/T_S$	$0.0497/T_S$	0
0.25	0.7071	$0.0332/T_S$	$0.0732/T_S$	0
0.30	0.6547	$0.0374/T_S$	$0.1035/T_S$	0.0119
0.35	0.6034	$0.0430/T_S$	$0.1341/T_S$	0.0579
0.40	0.5531	$0.0481/T_S$	$0.1603/T_S$	0.12

The impact of the gain α_D on the closed-loop performance of the current controller is illustrated in Table 5.1. For the gain that changes from 0.2 up to 0.4, the table comprises the corresponding values of the vector margin (VM), of the closed-loop bandwidths $f_{BW}(45°)$ and $f_{BW}(-3\ dB)$ and of the overshoot in the output response to the input step change. The data in the table are obtained analytically. The vector margin is obtained as the lowest value of $|1 + W_{OLG}(\exp(j\omega T_S))|$. The closed-loop bandwidth $f_{BW}(45°)$ is obtained as the frequency where the phase of the closed-loop transfer function (5.14) drops down to $-45°$, that is, to $-\pi/4$. The closed-loop bandwidth $f_{BW}(-3\ dB)$ is obtained as the frequency where the amplitude of the closed-loop transfer function (5.14) drops from 1 down to $1/\sqrt{2}$, that is, to $-3\ dB$.

According to results in Table 5.1, the closed-loop bandwidth of $f_{BW}(-3dB) = 0.1 \cdot f_S$ can be achieved with a very small overshoot and with the vector margin of VM $= 0.65$. Practical values of the gain α_D range from 0.25 up to 0.3. For $\alpha_D > 0.3$, the overshoot commences to rise, while the vector margin and the robustness of the system decay.

5.1.4   Disturbance Rejection

The closed-loop transfer function $W_{CL}(z)$ of (5.14) comprises only the real coefficients. Therefore, the reference current in d-axis would affect the current in the same axis while leaving the q-axis current undisturbed. This *decoupled* operation is obtained by the IMC controller (5.13). According to analytical predictions of Table 5.1, it is possible to reach the closed-loop bandwidth of 2 kHz in systems with the switching frequency of $f_{PWM} = 10$ kHz, where the sampling frequency is $f_S = 1/T_S = 20$ kHz.

In addition to suppressing the input disturbance (i.e., the step of the reference), it is also necessary to suppress the impact of the voltage disturbances, such as the back-electromotive forces, on the output current. If the current reference $i^*(z)$ in Fig. 5.2 is equal to zero, the output current $i^e(z)$ depends only on the voltage disturbance $e^e(z)$:

$$-W_{REG}(z) \cdot i^e(z) - z \cdot e^{j\frac{3}{2}\omega_e T_S} \cdot e^e(z) = \frac{1}{W_L(z)} i^e(z). \qquad (5.15)$$

From (5.15),

$$i^e(z) = -\frac{z \cdot e^{j\frac{3}{2}\omega_e T_S} W_L(z)}{1 + W_L(z) \cdot W_{REG}(z)} e^e(z) = -Y^e(z) \cdot e^e(z). \qquad (5.16)$$

By replacing (5.8) and (5.13) into (5.16),

$$Y^e(z) = \frac{z \cdot e^{j\frac{3}{2}\omega_e T_S} W_L(z)}{1 + W_L(z) \cdot W_{REG}(z)} = \frac{T_S}{L} \frac{z(z-1) \cdot e^{\frac{j}{2}\omega_e T_S}}{(z^2 - z + \alpha_D)(z \cdot e^{j\omega_e T_S} - e^{-\beta})} \qquad (5.17)$$

The pulse transfer function $Y^e(z)$ describes the output response to the voltage disturbance in z-domain. Namely, the output error caused by the voltage disturbance can be obtained from (5.16). It is of interest to understand the character of the output response to the voltage disturbance and to identify the parameters that have the largest impact on that response. To that purpose, it is assumed that the fundamental frequency ω_e is considerably smaller than the sampling frequency and that the factor $\exp(j\omega_e T_S)$ is close to one. Assuming at the same time that the voltage disturbance is a Heaviside step $e^e(z) = -E_d/(1-z^{-1})$, the output error becomes

$$i^e(z) = \frac{E_d \cdot z T_S}{z - 1} \frac{z(z-1)}{(z^2 - z + \alpha_D)(z - e^{-\beta})}. \qquad (5.18)$$

The response of the output current to the step voltage disturbance is given in Fig. 5.4. The waveform corresponds to the current error obtained with the voltage disturbance of $E_d = 1$ V. Namely, for the voltage step of 500 V, one would obtain current error that is 500 times larger than the waveform in Fig. 5.4. The peak value of the waveform in Fig. 5.4 is close to $T_S/L/\alpha_D$. For the given sampling time T_S and the load inductance, the peak value is defined by the gain α_D. In addition to the peak value, the waveform of Fig. 5.4 is characterized by its decay time. Disturbance rejection capability is larger if the decay time of the waveform in Fig. 5.4 is shorter. It is possible to quantify the disturbance rejection capability by measuring the surface below the curve of Fig. 5.4, namely, by determining the integral of the output error. The pulse transfer function of the output error integral $I(t)$ is given in (5.19). It is obtained by multiplying $i^e(z)$ by $1/(1-z^{-1})$. In order to calculate the surface below the error i^e, it is necessary to determine the final value of $I(t)$, that is, $I(\infty)$.

$$I(t) = \int_0^t i^e(\tau) \cdot d\tau,$$

$$Z(I(t)) = \frac{1}{1 - z^{-1}} i^e(z) = \frac{E_d \cdot z T_S}{z - 1} \frac{z^2}{L(z^2 - z + \alpha_D)(z - e^{-\beta})}. \qquad (5.19)$$

The final value $I(\infty)$ is obtained from the *final value theorem*:

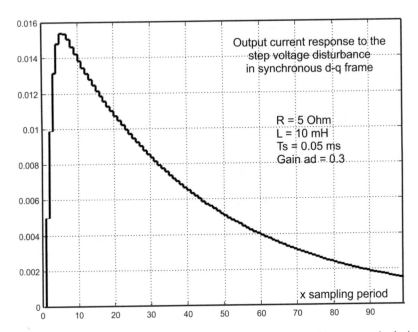

Fig. 5.4 Response of the output current to the voltage step disturbance. The response is obtained from (5.18) for the parameters R, L, T_S, and α_D that are listed within the figure

$$I(\infty) = \lim_{z \to 1} [(z - 1)I(z)],$$

$$I(\infty) = \frac{T_S}{L} \left. \frac{E_d \cdot z^3}{(z^2 - z + \alpha_D)(z - e^{-\beta})} \right|_{z \to 1} = \frac{T_S}{L} \frac{E_d}{\alpha_D(1 - e^{-\beta})}. \qquad (5.20)$$

With $\beta = RT_S/L < <1$, $1 - \exp(-\beta) \approx RT_S/L$. Therefore, the integral below the output current error can be obtained as

$$I(\infty) = \frac{E_d}{\alpha_D R}. \qquad (5.21)$$

Thus, the disturbance rejection depends on the load resistance R and on the gain α_D. The gain α_D is limited to the values given in Table 5.1, while the load resistance could be rather low, in particular in large power grid-side inverters and large power electrical drives.

5.2   Current Controller with Oversampling-Based Feedback

While the synchronous sampling with the feedback samples obtained at the center of the voltage pulses is simple to implement, the use of only one sample makes such feedback prone to the PWM noise and parasitic phenomena. In order to suppress the PWM noise and parasitic noise sources, it is necessary to use the feedback acquisition system described in Sect. 3.3. It is based on the oversampling, and it derives the feedback from the train of samples acquired within the past PWM period. This one-PWM-period averaging improves the robustness of the controller against the noise in the feedback path. All the PWM-related noise is removed, while the robustness against spurious and parasitic noise sources is improved. At the same time, the one-PWM-period averaging introduces the transport delays which have to be taken into considerations in designing the current controller structure and setting the parameters.

5.2.1   The Pulse Transfer Function of the Feedback Path

The block diagram of the d-q frame digital current controller with one-PWM-period averaging is shown in Fig. 5.5. The feedback signal i^{FB} at instant $t = nT_S$ is obtained by calculating the average value of N_{OV} samples acquired on the interval $[(n-2)T_S .. nT_S]$. Corresponding time schedule of control tasks is given in Fig. 5.6. The voltage reference u_n^* is calculated in interrupt routine triggered at nT_S, and it affects the average voltage on the interval $[(n+1)T_S .. (n+2)T_S]$. Therefore, the pulse transfer function of the load $W_L(z)$ is the same as the one in (5.8).

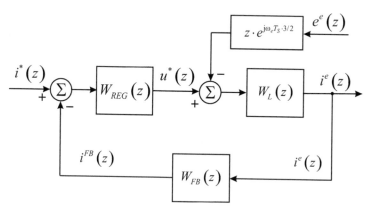

Fig. 5.5 Block diagram of the current controller with one-PWM-period averaging oversampling-based feedback

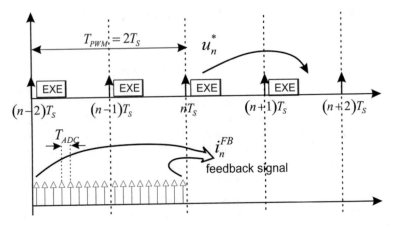

Fig. 5.6 Time schedule of the feedback acquisition, computation, and pulse width modulation of the digital current controller with one-PWM-period averaging in the feedback path

In Fig. 5.6, the oversampling period T_{ADC} is equal to T_{PWM}/N_{OV}, where N_{OV} is the number of samples in each PWM period. The value i^{FB} at instant nT_S represents the average of the current over the interval $[(n-2)T_S .. nT_S]$. It can be expressed in terms of the samples of the output current i^e in d-q frame. These samples are $i_{n-2} = i^e(nT_S-2T_S)$, $i_{n-1} = i^e(nT_S-T_S)$, and $i_n = i^e(nT_S)$. The feedback signal can be expressed as $i^{FB}{}_n = (i_{n-2} + 2i_{n-1} + i_n)/4$. Therefore, transfer function in the feedback path is given by $W_{FB}(z)$ in (5.21). More detailed analysis and developments, related to the one-PWM-period averaging, are given in Sect. 3.3:

$$W_{FB}(z) = \frac{i^{FB}(z)}{i(z)} = \frac{z^2 + 2 \cdot z + 1}{4 \cdot z^2} \tag{5.22}$$

5.2.2   Design of the Controller Structure

The block diagram of the current controller is given in Fig. 5.5. Within the figure, the input disturbance is the current reference $i^*(z) = i^*{}_d(z) + ji^*{}_q(z)$, the voltage disturbance is $e^e(z) = e^e{}_d(z) + je^e{}_q(z)$, while the output current is $i^e(z) = i^e{}_d(z) + ji^e{}_q(z)$. The pulse transfer function of the load $W_L(z)$ is given in (5.8). The feedback chain is described by the pulse transfer function (5.22), which includes the relevant transport delays. Delays within the PWM modulator and the voltage actuator are included in $W_L(z)$. The IMC design consists in inverting dynamics of $W_L(z)$ and multiplying the result by the transfer function of an integrator, namely, with the pulse transfer function $\alpha_D/(1 - z^{-1}) = \alpha_D \cdot z/(z - 1)$, where α_D is an adjustable gain. An attempt to invert the pulse transfer function of the load $W_L(z)$ and to multiply the outcome by the factor $\alpha_D \cdot z/(z-1)$ results in the pulse transfer function $W_{REGX}(z)$ of (5.10), which cannot be implemented, since it implies prediction. In order to obtain the practical current controller, the pulse transfer function $W_{REGX}(z)$ has to be divided by z^2, thus resulting in

$$W_{REG}(z) = \frac{\alpha_D L}{T_S} \frac{e^{j\omega_e T_S} \left(z \cdot e^{j\omega_e T_S} - e^{-\beta} \right)}{z - 1} \tag{5.23}$$

With (5.23), the product of the two pulse transfer functions in the direct path becomes

$$W_{REG\_L}(z) = W_{REG}(z) W_L(z) = \frac{\alpha_D}{z(z-1)}. \tag{5.24}$$

The closed-loop transfer function $W_{CL}(z)$ for the system of Fig. 5.5 is

$$W_{CL}(z) = \frac{W_{REG\_L}(z)}{1 + W_{REG\_L}(z) \cdot W_{FB}(z)}. \tag{5.25}$$

By introducing the results (5.22) and (5.24) into (5.25), the closed-loop transfer function W_{CL} is

$$W_{CL}(z) = \frac{4\alpha_D z^2}{4z^4 - 4z^3 + \alpha_D z^2 + 2\alpha_D z + \alpha_D}. \tag{5.26}$$

Compared to the closed-loop transfer function (5.14), W_{CL} of (5.26) has the polynomial of the fourth order in denominator. This implies the four closed-loop poles of the pulse transfer function $W_{CL}(z)$. Additional two poles are introduced due to the transport delay within the feedback path, introduced by $W_{FB}(z)$. In (5.26), both numerator and denominator have real coefficients. With $W_{CL}(z) = i^e(z)/i^*(z) = (i_d^e + ji_q^e)/(i_d^* + ji_q^*)$, this means that the output current in d-axis does not depend on the current reference in q-axis and vice versa. Namely, the current controller ensures decoupled operation.

5.2.3   Parameter Setting

Due to delays introduced by one-PWM-period feedback averaging and the corresponding pulse transfer function $W_{FB}(z)$, the characteristic polynomial $f(z) = 4z^4 - 4z^3 + \alpha_D z^2 + 2\alpha_D z + \alpha_D$ has four roots. Even with very low gain α_D, at least two of the poles are conjugate complex. Hence, the feedback delay introduced by the proposed feedback averaging produces a considerable reduction of the closed-loop performances. With $\alpha_D = 0.1$, the closed-loop poles are $z_1 = 0.8660$, $z_2 = 0.4642$, and $z_{3/4} = -0.1651 \pm j \cdot 0.1869$. The poles in z-domain and the corresponding poles in s-domain are correlated by $z = \exp(sT_S)$. With the

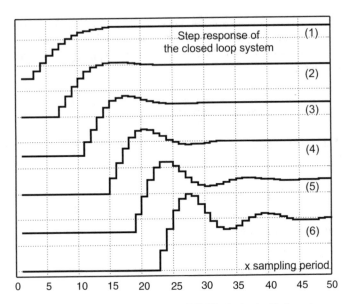

Fig. 5.7 Step response of the closed-loop system of (5.26) obtained with the gain α_D equal to 0.15 (trace 1), 0.2 (trace 2), 0.25 (trace 3), 0.3 (trace 4), 0.35 (trace 5), and 0.4 (trace 6).

Table 5.2 Closed-loop performance with WCL of (5.26)

Gain α_D	VM	$f_{BW}(45°)$	$f_{BW}(-3\ dB)$	Overshoot
0.15	0.7218	$0.0227/T_S$	$0.0434/T_S$	0.0001
0.20	0.6423	$0.0298/T_S$	$0.0708/T_S$	0.0445
0.25	0.5664	$0.0370/T_S$	$0.0939/T_S$	0.1406
0.30	0.4935	$0.0442/T_S$	$0.1110/T_S$	0.2510
0.35	0.4230	$0.0513/T_S$	$0.1245/T_S$	0.3558
0.4	0.3548	$0.0585/T_S$	$0.1349/T_S$	0.4840

sampling time of $T_S = 50$ µs, corresponding s-domain poles are $s_1 = -2878$, $s_2 = -15{,}348$, and $s_{3/4} = -27775 \pm j \cdot 0.45885$ [rad/s]. The pole $s_1 = -2878$ is dominant, and it can contribute to the step response without an overshoot, notwithstanding the conjugate complex poles $s_{3/4}$. The step response of the closed-loop system is plotted in Fig. 5.7 for the gain α_D ranging from 0.15 up to 0.4.

The impact of the gain α_D on the closed-loop transfer function (5.26) and the closed-loop performance of the current controller is illustrated in Table 5.2. For the gain α_D that changes from 0.15 up to 0.4, the table comprises the corresponding values of the vector margin (VM), which provides the information on the robustness, that is, on the capability of the controller to maintain the response quality in the presence of parameter changes. In addition, the table comprises of the closed-loop bandwidths $f_{BW}(45°)$ and $f_{BW}(-3\ dB)$ and also of the overshoot in the output response to the input step change. The data of the Table are obtained analytically. The vector margin is obtained as the lowest value of $|1 + W_{OLG}(\exp(j\omega T_S))|$, where

W_{OLG} is the open-loop gain, that is, the product of W_{REG_L} of (5.24) and W_{FB} of (5.22). The closed-loop bandwidth $f_{BW}(45°)$ is obtained as the frequency where the phase of W_{CL} drops down to $-\pi/4$. The closed-loop bandwidth $f_{BW}(-3\ dB)$ is obtained as the frequency where the amplitude of W_{CL} drops down to $-3\ dB$. The overshoot is obtained by simulating the input step response.

According to results in Table 5.2, the closed-loop bandwidth of $f_{BW}(-3dB) = 0.07{\cdot}f_S$ can be achieved with an overshoot of 4.4% and with the vector margin of $VM = 0.64$. For $\alpha_D > 0.25$, the overshoot assumes considerable values while the vector margin decays. Thus, the range of applicable gains is up to 0.2.

The current controller with one-PWM-period averaging is shown in Fig. 5.5. While the feedback averaging contributes to the suppression of the PWM-related noise, the associated delays reduce the closed-loop bandwidth and reduce the range of applicable gains. With synchronous center-pulse sampling, the gain α_D reaches 0.3, while the closed-loop bandwidth reaches $f_{BW}(-3dB) = 0.1{\cdot}f_S$. With one-PWM-period averaging, the gain α_D reaches 0.2, while the closed-loop bandwidth reaches $f_{BW}(-3dB) = 0.07{\cdot}f_S$.

5.2.4   Disturbance Rejection

Disturbance rejection capability of the current controller with one-PWM-period feedback averaging is obtained from the pulse transfer function:

$$Y^e(z) = \left.\frac{i^e(z)}{-e^e(z)}\right|_{i^*(z)=0}. \tag{5.27}$$

From the block diagram in Fig. 5.5,

$$Y^e(z) = \frac{z \cdot e^{j\frac{3}{2}\omega_e T_S} W_L(z)}{1 + W_L(z) \cdot W_{REG}(z) \cdot W_{FB}(z)} = \frac{z \cdot e^{j\frac{3}{2}\omega_e T_S} W_L(z)}{1 + W_{REG\_L}(z) \cdot W_{FB}(z)}. \tag{5.28}$$

By introducing $W_{FB}(z)$ from (5.22), $W_{REG\_L}(z)$ from (5.24), and $W_L(z)$ from (5.8), the pulse transfer function Y^e becomes

$$Y^e(z) = \frac{T_S}{L} \frac{4 \cdot z^3(z-1) \cdot e^{j\frac{1}{2}\omega_e T_S}}{(4z^4 - 4z^3 + \alpha_D z^2 + 2\alpha_D z + \alpha_D)(z \cdot e^{j\omega_e T_S} - e^{-\beta})} \tag{5.29}$$

Assuming that the voltage disturbance is a Heaviside step, that is, $e^e(z) = -E_d/(1-z^{-1})$, the integral of the output current error of Fig. 5.4 becomes

$$I(\infty) = \lim_{z \to 1}[(z-1)I(z)] = \lim_{z \to 1}\left[(z-1) \cdot \frac{z}{z-1} \cdot \frac{E_d \cdot z}{z-1} \cdot Y^e(z)\right],$$

$$I(\infty) = \left.\frac{T_S}{L}\frac{E_d}{\alpha_D(1-e^{-\beta})}\right| = \frac{E_d}{\alpha_D R}. \tag{5.30}$$

Thus, the expression (5.30) for the integral of the current error encountered with the voltage step change is the same as the one of (5.22). The gain α_D is lower with one-PWM-period averaging, resulting in a lower disturbance rejection capability.

5.3 Current Controllers with Series Compensator

The limiting factors of the closed-loop performance are the transport delays in the feedback acquisition path and the transport delays incurred within the process of the pulse width modulation. The effect of delays can be reduced by introducing the series compensator with differential action. A differential series compensator is known to improve the phase characteristic, increase the damping factor, and reduce the overshoot. Both current controllers discussed in this chapter are extended in order to reduce the effects of the transport delays. Their closed-loop performance with the series differential compensator is evaluated analytically and by means of computer simulations.

5.3.1 Synchronous Sampling with Series Compensator

The block diagram of the d-q frame digital current controller with synchronous, center-pulse sampling and with the differential series compensator is shown in Fig. 5.8. The current controller $W_{\text{REG}}(z)$ of (5.13) is designed to cancel out the undesired dynamics of the load. The elements of the load dynamics that cannot be canceled include the transport delays. They affect the phase characteristic of the closed-loop gain, they reduce the damping of the conjugate complex poles, and they increase the overshoot. Some effects of the transport delays can be reduced by

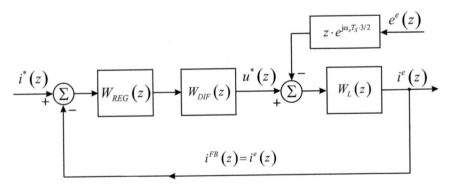

Fig. 5.8 IMC-based digital current controller with synchronous, center-pulse sampling and with the differential series compensator

introduction of the series differential compensator. The pulse transfer function of the series compensator is given in (5.31). With $d = 0$, it reduces to the unity gain.

$$W_{\text{DIF}}(z) = 1 + d\frac{z-1}{z} = \frac{(1+d)z - d}{z}. \tag{5.31}$$

Considering the current controller design $W_{\text{REG}}(z)$ of (5.13) and multiplying the pulse transfer function of the controller by the series compensator $W_{\text{DIF}}(z)$, one obtains

$$W_{\text{REG}}^{\text{NEW}}(z) = W_{\text{REG}}(z) \cdot W_{\text{DIF}}(z). \tag{5.32}$$

Replacing $W_{\text{REG}}(z)$ from (5.13) and $W_{\text{DIF}}(z)$ from (5.31),

$$
\begin{aligned}
W_{\text{REG}}^{\text{NEW}}(z) &= \left(1 + d\frac{z-1}{z}\right) \cdot \left(\frac{\alpha_D L}{T_S} \cdot \frac{e^{j\omega_e T_S}\left(z \cdot e^{j\omega_e T_S} - e^{-\beta}\right)}{z-1}\right) = \\
&= \frac{\alpha_D L}{T_S} e^{j\omega_e T_S} \frac{[(d+1)z - d]\left(z \cdot e^{j\omega_e T_S} - e^{-\beta}\right)}{z(z-1)}.
\end{aligned}
\tag{5.33}
$$

The closed-loop pulse transfer function of the system in Fig. 5.8 is

$$W_{\text{CL}}(z) = \frac{W_{\text{REG}}^{\text{NEW}}(z) \cdot W_L(z)}{1 + W_{\text{REG}}^{\text{NEW}}(z) \cdot W_L(z)}. \tag{5.34}$$

Introducing W_L of (5.8) and the enhanced controller from (5.33) into the expression (5.34), the closed-loop pulse transfer function becomes

$$W_{\text{CL}}(z) = \frac{\alpha_D(1+d)z - \alpha_D d}{z^3 - z^2 + \alpha_D(1+d)z - \alpha_D d}. \tag{5.35}$$

Compared to the closed-loop transfer function of (5.14), obtained without the series compensator, the pulse transfer function of (5.35) has one more pole, but it also has one zero in denominator. The zero is determined by the parameter d, and it improves the phase characteristic of the closed-loop system.

The impact of the differential gain on the closed-loop performance of the current controller is studied by observing the vector margin, the closed-loop bandwidth, and the overshoot for $\alpha_D = 0.35$ and for the gain d changing from 0 up to 0.25. The Matlab code used to calculate the closed-loop performance is given in Table 5.3. The results are listed in Table 5.4. While the gain d rises from 0 up to 0.25, the overshoot reduces to zero, the closed-loop bandwidth increases, while the vector margin remains close to 0.6. The impact of the series differential compensator on the closed-loop performance of the d-q frame current controller with synchronous sampling is positive, but the consequential increase of the closed-loop bandwidth is relatively modest. The values of α_D and d in the table are relative, and they do not depend on the motor or the grid parameters.

Table 5.3 Matlab code which calculates the vector margin, the closed-loop bandwidth, and the overshoot in the step response of WCL(z) of (5.35)

```
a = 0.35;                          % Set the parameter αD
d = 0.1;                           % Set the differential gain
num = [a*(1+d)-a*d];               % Numerator of WCL of (5.34)
den = [1-1 a*(1+d)-a*d];           % Denominator of WCL of (5.34)
iout = dstep(num,den,50);          % Input step response

for jj=1:1000,W(jj)=jj*50;end;     % Generating the vector W [0 .. Wmax]
z = exp(1i*W*T);                   % Generating the vector z
WPP = (a*(1+d)*z-a*d);             % Calculating the open loop gain for
WPP = WPP./z./z./(z-1);            % W sweeping from 0 to Wmax
VM = min(abs(1+WPP));              % Calculating the vector margin
Preb = max(iout);                  % Calculating the overshoot

[Mag, Pha, W] = dbode(num,den,0.00005,W); % Getting the amplitude and
                                          % frequency characteristic
f45 = W(min(find(Pha< -45)))/2/pi;        % from Bode-plot. Getting
fbw = W(min(find(Mag<0.707)))/2/pi;       % the bandwidth

disp('a  d  VM  f45  f3db  preb ')        % Printing the outputs
[a  d  VM  f45*0.00005  fbw*0.00005 Preb]
```

Table 5.4 The impact of the series differential compensator on WCL of (5.26)

Gain α_D	Gain d	VM	$f_{BW}(45°)$	$f_{BW}(-3\text{ dB})$	Overshoot
0.35	0	0.6034	$0.0429/T_S$	$0.1340/T_S$	0.0579
0.35	0.05	0.6076	$0.0430/T_S$	$0.1360/T_S$	0.0451
0.35	0.1	0.6104	$0.0433/T_S$	$0.1384/T_S$	0.0323
0.35	0.15	0.6115	$0.0437/T_S$	$0.1416/T_S$	0.0197
0.35	0.2	0.6109	$0.0441/T_S$	$0.1456/T_S$	0.0072
0.35	0.25	0.6085	$0.0441/T_S$	$0.1508/T_S$	0.0000

5.3.2   One-PWM-Period Averaging with Series Compensator

The block diagram of the d-q frame digital current controller with one-PWM-period averaging, oversampling-based feedback acquisition, and the differential series compensator is shown in Fig. 5.9. The pulse transfer function of the current controller cancels out the undesired dynamics of the load, while the pulse transfer function W_{FB} provides the model of the feedback acquisition subsystem. The transport delays affect the phase characteristic of the closed-loop gain, reduce the damping, and increase the overshoot. The series differential compensator is introduced to reduce the negative effects of the delays. The pulse transfer function of the series compensator is given in (5.31).

The product of the current controller pulse transfer function $W_{REG}(z)$ and the pulse transfer function of the series differential compensator $W_{DIF}(z)$ of (5.31) is

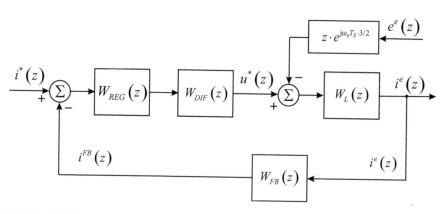

Fig. 5.9 IMC-based digital current controller with one-PWM-period feedback averaging and with the differential series compensator

$$
\begin{aligned}
W_{REG}^{NEW}(z) &= \left(1 + d\,\frac{z-1}{z}\right) \cdot \left(\frac{\alpha_D L}{T_S} \cdot \frac{e^{j\omega_e T_S}\left(z \cdot e^{j\omega_e T_S} - e^{-\beta}\right)}{z-1}\right) \\
&= \frac{\alpha_D L}{T_S}\,e^{j\omega_e T_S}\,\frac{[(d+1)z - d]\left(z \cdot e^{j\omega_e T_S} - e^{-\beta}\right)}{z(z-1)}.
\end{aligned}
\tag{5.36}
$$

The closed-loop pulse transfer function of the system in Fig. 5.9 is

$$
W_{CL}(z) = \frac{W_{REG}^{NEW}(z) \cdot W_L(z)}{1 + W_{REG}^{NEW}(z) \cdot W_L(z) \cdot W_{FB}(z)}.
\tag{5.37}
$$

Introducing W_L and W_{FB} into (5.37), the closed-loop pulse transfer function becomes

$$
W_{CL}(z) = \frac{4\alpha_D(1+d)z^3 - 4\alpha_D dz^2}{4z^5 - 4z^4 + \alpha_D(1+d)z^3 + \alpha_D(2+d)z^2 + \alpha_D(1-d)z - \alpha_D d}.
\tag{5.38}
$$

Modified closed-loop transfer function, obtained with the series compensator, has one pole more than the original closed-loop transfer function of (5.26), but it also has one finite zero in denominator. The zero is determined by the parameter d, and it improves the phase characteristic of the closed-loop system.

The Matlab code used to calculate the closed-loop performance is given in Table 5.5. The code calculates the vector margin, the closed-loop bandwidth, and the overshoot for $\alpha_D = 0.25$ and for the gain d changing from 0 up to 1. The results are listed in Table 5.6. While the gain d rises from 0 up to 1, the overshoot reduces to zero, the closed-loop bandwidth increases, while the vector margin remains close to 0.58. The impact of the series differential compensator on the closed-loop performance of the d-q frame current controller with synchronous sampling is positive, and it is considerably more emphasized than in the previous case (Sect. 5.3.1). The values of α_D and d in the table are relative, and they do not depend on the motor or the grid parameters.

Table 5.5 Matlab code which calculates the vector margin, the closed-loop bandwidth, and the overshoot in the step response of WCL(z) of (5.38)

```
a = 0.25;                                    % Set the parameter αD
d = 0.0;                                      % Set the differential gain
num =[4*a*(1+d)-4*a*d 0 0];                   % Numerator of WCL of (5.37)
den =[4-4 a*(1+d)+a*(2+d)+a*(1-d)-a*d]; % Denominator of WCL of (5.37)
iout = dstep(num,den,50);                    % Input step response

for jj=1:1000, W(jj)=jj*50; end;             % Generating the vector W [0 .. Wmax]
z = exp(1i*W*T);                              % Generating the vector z
WPP = (a*(1+d)*z-a*d);                        % Calculating the open loop gain for
WPP = WPP./z./z./(z-1);                       % W sweeping from 0 to Wmax
WPP = WPP.*(1+2./z+1./z./z)/4;
VM = min(abs(1+WPP));                         % Calculating the vector margin
Preb = max(iout);                            % Calculating the overshoot
[Mag,Pha,W] = dbode(num,den,0.00005,W); % Getting the amplitude and
                                              % frequency characteristic
f45 = W(min(find(Pha<-45)))/2/pi;            % from Bode-plot. Getting
fbw = W(min(find(Mag<0.707)))/2/pi;          % the bandwidth
disp('a  d  VM  f45  f3db  preb ')           % Printing the outputs
[a  d  VM  f45*0.00005 fbw*0.00005 Preb]
```

Table 5.6 The impact of the series differential compensator on W_{CL} of (5.28)

Gain α_D	Gain d	VM	$f_{BW}(45°)$	$f_{BW}(-3\,dB)$	Overshoot
0.25	0	0.5664	$0.0370/T_S$	$0.0939/T_S$	0.1406
0.25	0.2	0.5869	$0.0382/T_S$	$0.0975/T_S$	0.1041
0.25	0.4	0.6006	$0.0394/T_S$	$0.1031/T_S$	0.0685
0.25	0.6	0.6054	$0.0410/T_S$	$0.1118/T_S$	0.0394
0.25	0.8	0.6003	$0.0434/T_S$	$0.1261/T_S$	0.0126
0.25	1	0.5863	$0.0458/T_S$	$0.1448/T_S$	0.0000

5.4   Experimental Runs with IMC-Based Controllers

In this chapter, the analytical considerations are focused on designing and tuning the d-q frame digital current controllers that are based on the internal model control. The controller structure is designed by attempting to cancel out the undesired dynamics of the load. The transfer function of the controller is obtained by attempting an inversion of the transfer function of the load. In addition, the chapter discusses the parameter setting, the disturbance rejection capability, and the possibility to reduce the impact of transport delays by adding a series compensator of differential nature. The closed-loop performance of several controllers is probed analytically and by means of the computer simulation. In this section, the comparison of the closed-loop performances is aided by the experimental results.

The computer simulations and the experimental results are often different due to the non-modeled dynamics that affects the actual load, while they are not taken into account in the simulation models. An example of such phenomena is the parasitic capacitance of the windings, frequency dependence of the load parameters, non-modeled iron losses, and similar. Additional reason for the differences between the actual experimental results and the corresponding simulation runs is the noise that affects the measurements. The noise can be related to the inverter switching and the pulse width modulation, but it can also come from other sources such as the switch mode power supply, the ac grid, and others.

The experimental setup includes a brushless dc motor and a three-phase inverter. The relevant parameters of the experimental setup are listed in the subsequent section.

5.4.1   Parameters of the Experimental Setup

The three-phase inverter of the experimental setup is supplied from a dc voltage source that provides $E_{DC} = 520$ V. Within the experimental runs presented in this section, the PWM frequency is equal to 7.812 kHz. Thus, the sampling period is equal to $T_S = T_{PWM}/2 = 64$ μs. The DSP controller used to implement the devised algorithms is TMS320F28335. The DSP device is equipped with a $N = 12$-bit ADC peripheral. The anti-aliasing filters are passive RC filters with the time constant of 5 μs. The oversampling interval is set to $T_{ADC} = 4$ μs. Within each PWM period, the feedback acquisition system collects $N_{OV} = T_{PWM}/T_{ADC} = 32$ samples of the feedback currents. The samples are acquired and collected automatically, by means of a dedicated DMA machine that reads the DSP peripherals and places the samples into the RAM memory of the controller.

The PWM inverter uses the IGBT power switches, and it has a lockout time of 3 μs. The lockout time is compensated by correcting the modulation signals in accordance with the sign of the output currents. The peak output current is 45 A. The three-phase load used in experimental runs is a brushless dc motor with the rated current of 7.3 A, with the phase resistance of $R = 0.47$ Ω, with the phase inductance of 3.4 mH, and with the back-electromotive force constant of 0.687 Vpeak/(rad/s).

5.4.2   Experimental Results

The experiments were performed by introducing the step change of the current reference for the q-axis. The traces in Figs. 5.10 and 5.11 are obtained with the three-phase brushless dc motor as the load. The traces in Fig. 5.12 are obtained at very high output frequency f_e and with the load that comprises three star-connected inductances.

In Fig. 5.10, the experimental traces represent the q-axis current obtained at the output of the feedback acquisition chain. The output frequency is relatively low, while the closed-loop bandwidth is $f_{BW} = 0.1/T_S$. The reference current for the q-axis is changed in a stepwise manner. The upper trace corresponds to synchronous

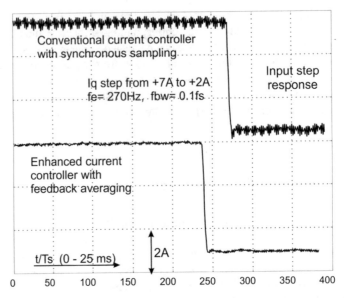

Fig. 5.10 Experimental traces of the q-axis current step response at relatively low output frequency f_e and with the closed-loop bandwidth $f_{BW} = 0.1/T_S$. The upper trace corresponds to synchronous center-pulse sampling feedback acquisition and the controller without the series compensator. The lower trace corresponds to the one-PWM-period feedback averaging and the enhanced controller which includes the feedback averaging

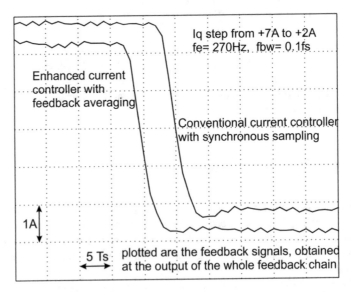

Fig. 5.11 The traces from the previous figure are enlarged for the ease of comparison

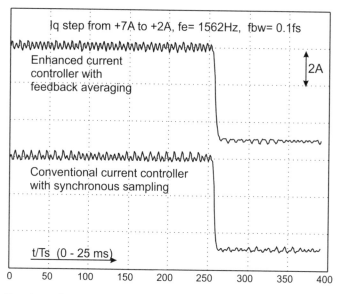

Fig. 5.12 Experimental traces of the q-axis current step response at relatively high output frequency f_e, with the closed-loop bandwidth $f_{BW} = 0.1/T_S$ and with the load that includes three star-connected inductances. The upper trace corresponds to synchronous center-pulse sampling feedback acquisition and the controller without the series compensator. The lower trace corresponds to the one-PWM-period feedback averaging and the enhanced controller which includes the feedback averaging

center-pulse sampling feedback acquisition and the controller without the series compensator. The lower trace corresponds to the one-PWM-period feedback averaging and the enhanced controller which includes the feedback averaging. The upper trace has a considerable amount of noise. The noise is present even within the feedback signal due to the fact that the feedback acquisition system acquires one sample in each $T_S = T_{PWM}/2$ period. In vicinity of power electronics devices with high dV/dt and di/dt values, spurious effects and non-modeled parasitic capacitances contribute to the noise. The lower trace is obtained with one-PWM-period averaging, where the feedback signal is obtained as an average of $N_{OV} = 32$ samples, collected within the past PWM period. Therefore, the amount of noise is considerably lower.

In order to compare the rise time and the overshoot of the step response, the waveforms of Fig. 5.10 are enlarged and focused on the instant of the step transient (Fig. 5.11). One of the waveforms is obtained from the noise-sensitive synchronous sampling-based controllers where the feedback samples are acquired at the center of the voltage pulses. The second waveform is obtained with one-PWM-period feedback averaging, which filters out the noise, but it also introduces the transport delay. The current controller is enhanced by introducing the series differential compensator. From Fig. 5.11, it is evident that the series compensator reduces the impact of the transport delay and provides the step response waveform comparable to the one obtained with synchronous sampling.

When the frequency f_e of the output current becomes larger, it is more difficult to control the current and provide the error-free operation. In order to verify the capability of the current controller to maintain an error-free control even at very high output frequencies, the experimental runs were performed with very large output frequency (Fig. 5.12). The load was replaced by the three star-connected inductances. The experimental traces demonstrate the capability of the IMC-based controllers to maintain the response character even at very high output frequencies.

Question (5.1) For the discrete-time current controller of Fig. 5.2, with the IMC-based controller of (5.13), determine the maximum gain α_D which results in the closed-loop poles in s-domain with the damping of $\xi = 0$. By increasing the gain gradually, find the approximate value of α_D which gives $\xi = 0.5$.

Answer (5.1) The closed-loop controller is designed in (5.12):

$$W_{REG}(z) = \frac{\alpha_D L}{T_S} \frac{e^{j\omega_e T_S}\left(z \cdot e^{j\omega_e T_S} - e^{-\beta}\right)}{z - 1}$$

The open-loop gain becomes

$$W_{OLG}(z) = W_{REG}(z)W_L(z) = \frac{\alpha_D}{z(z-1)}.$$

Thus, the closed-loop transfer function becomes

$$W_{CL}(z) = \frac{i^e(z)}{i^*(z)}\bigg|_{e^e=0} = \frac{W_{OLG}(z)}{1 + W_{OLG}(z)} = \frac{\alpha_D}{z^2 - z + \alpha_D}.$$

The characteristic polynomial $f(z) = z^2 - z + \alpha_D$ has two poles and only one gain. Relation between the poles in s-domain and the poles in z-domain is

$$z_1 = e^{s_1 T}, \quad z_2 = e^{s_2 T}$$

For low values of the gain ($\alpha_D = 0.1$), the z-domain poles are both real (0.8873 and 0.1127), as well as the poles in s-domain. When the gain is increased ($\alpha_D = 0.2$), the z-domain poles remain real (0.7236 and 0.2764), and they are coming closer to each other. For $4\alpha_D = 1$, the poles in z-domain are both equal to 0.5. At this point, the s-domain poles are both real and equal. Any further increase in the gain results in conjugate complex poles in both s-domain and z-domain. With the gain of 0.3, the z-domain poles are $z_{1/2} = 0.5000 \pm j0.2236$. With the sampling time T, these poles correspond to s-domain poles $s_{1/2} = (1/T)(-0.6020 \pm j0.4205)$, with the damping factor $\xi = 0.602/|-0.6020 \pm j0.4205| = 0.819$. The damping factor drops down to 0.5 for $\alpha_D = 0.438$.

Question (5.2) Considering the system of Question 5.1, and using the computer math tools, determine the maximum gain α_D which results in the step response with an overshoot lower than 25%. Determine the vector margin for this gain.

Answer (5.2) Analytical solution of the problem is quite involved. Therefore, it is of interest to use the math tools in order to find the solution. One of them is Matlab. The following sequence of commands provides the information on the overshoot that is obtain for the given gain:

```
>> Gain = 0.25;
>> Response = dstep(Gain,[1 -1 Gain]);
>> Overshoot = max(Response) -1;
```

In order to find the largest gain with an overshoot lower than 25%, it is possible to organize the search where the gain is gradually increased:

```
>> Overshoot = 0; Gain = 0.2; % Initial values
>> while Overshoot < 0.25;
>>    Gain = Gain + 0.00001;
>>    Overshoot = max(dstep(G,[1 -1 G]))-1;
>> end;
```

The above script results in gain $= \alpha_D = 0.5$.

At this point, it is necessary to calculate the *vector margin* (VM). The vector margin can be derived from the open-loop transfer function of (5.13), $W_{OLG}(z) = \alpha_D/z/(z-1)$. The stability margin can be judged from the lowest value of $|1 + W_{OLG}(z)|$. Assuming that the frequency ω sweeps from $-\infty$ to $+ -\infty$ and taking into account the relation $z = \exp(j\omega T_S)$, the vector margin is obtained as the lowest value of $|1 + W_{OLG}(z)|$, obtained when the graph $W_{OLG}(\exp(j\omega T_S))$ comes close to the point $(-1, 0)$ of the z-plane. The vector margin can be found by running the script given below. The vector margin found is VM $= 0.4551$:

```
>> Gain = 0.5;
>> fsampling = 10000;
>> VM = 1;
>> for j = 1:fsampling/2,
>>    Wrads = 1i*j*2*pi;
>>    z=exp(Wrads/fsampling);
>>    WOLG = Gain/z/(z-1);
>>    WW = abs(1+WOLG);
>>    if WW < VM, VM = WW; end;
>> end
```

Question (5.3) For the discrete-time current controller with one-PWM-period feed-back averaging, described by the closed-loop transfer function of (5.26), and using the computer math tools, determine the maximum gain α_D which results in the step response with an overshoot lower than 25%.

Answer (5.3) The closed-loop transfer function of the given controller is given by

$$W_{CL}(z) = \frac{4\alpha_D z^2}{4z^4 - 4z^3 + \alpha_D z^2 + 2\alpha_D z + \alpha_D}.$$

The characteristic polynomial has four poles and only one gain. The following sequence of commands provides the information on the overshoot that is obtain for the given gain:

```
>> Gain = 0.25;
>> Response = dstep(4*Gain, [4 -4 Gain 2*Gain Gain]);
>> Overshoot = max(Response) -1;
```

In order to find the largest gain with an overshoot lower than 25%, it is possible to organize the search where the gain is gradually increased:

```
>> Overshoot = 0; Gain = 0.2; % Initial values
>> while Overshoot < 0.25;
>>    Gain = Gain + 0.00001;
>>    Overshoot = max(dstep(4*Gain, [4 -4 Gain 2*Gain Gain]))-1;
>> end;
```

The above script results in gain $= \alpha_D = 0.3$.

Question (5.4) Considering the system of Question 5.3, and using the computer math tools, determine the vector margin for the given gain.

Answer (5.4) The vector margin is calculated in a manner similar to this of Question 5.2. With the frequency sweeping from zero up to one half of the sampling frequency, and taking into account the relation $z = \exp(j\omega T_S)$, the vector margin is the lowest value of $|1 + W_{OLG}(z)|$, obtained when the graph $W_{OLG}(\exp(j\omega T_S))$ comes close to the point $(-1, 0)$ of the z-plane. The open-loop gain $W_{OLG}(z)$ is

$$W_{OLG}(z) = W_{REG\_L}(z) = W_{REG}(z)W_L(z)$$
$$= \frac{\alpha_D}{z(z-1)} \cdot \frac{z^2 + 2 \cdot z + 1}{4 \cdot z^2}$$

The vector margin can be found by running the script given below. The vector margin found is $VM = 0.4935$, which is in good agreement with the results given in Table 5.2:

```
>> Gain = 0.3;
>> fsampling = 10000;
>> VM = 1;
>> for j = 1:fsampling/2,
>>    Wrads = 1i*j*2*pi;
```

```
>>     z=exp(Wrads/fsampling);
>>     WOLG = Gain*(z*z+2*z+1)/z/z/z/(z-1)/4;
>>     WW = abs(1+WOLG);
>>     if WW < VM, VM = WW; end;
>> end
```

Chapter 6
Scheduling of the Control Tasks

In this Chapter, the closed loop performance of digital current controllers is improved by introducing advanced scheduling of the control tasks. The Chapter provides the relevant analysis, an insight into the available scheduling options; and it discusses the scheduling schemes that have the potential of reducing the transport delay within digital current controllers. For the most promising scheduling schemes, the analysis provides the design and parameter setting along with an evaluation and comparison to the previous solutions.

The basic control tasks include the oversampling process, the averaging of the acquired feedback samples, the execution of the control algorithm, and the implementation of the pulse width modulation of the voltage impulses. The organization and scheduling of the control interrupt and the synchronization of the tasks can be organized in the way that reduces the transport delays and improves the closed-loop performance. An improved current controller scheduling can provide a higher closed-loop bandwidth, better robustness, and an improved disturbance rejection. In conjunction with a series compensator of differential character, the closed-loop bandwidth can be increased from $0.1/T_S$ up to $0.17/T_S$, doubling at the same time the disturbance rejection capability, achieving the step response with a negligible overshoot, while maintaining the robustness with a vector margin of 0.65. The chapter comprises analytical considerations, design of the scheduling scheme, calculation of the relevant pulse transfer functions, parameter setting, analytical evaluation of the closed-loop performances, and experimental verification.

6.1 Scheduling Schemes

Digital current controller is commonly implemented on digital signal processors, the microprocessors capable of performing numerical tasks rather quickly. Contemporary digital signal processors (DSP) are capable of executing one floating-point operation in several nanoseconds. Some DSP devices are integrated with peripheral

© Springer International Publishing AG, part of Springer Nature 2018
S. N. Vukosavic, *Grid-Side Converters Control and Design*, Power Electronics
and Power Systems, https://doi.org/10.1007/978-3-319-73278-7_6

devices, such as the A/D converters, PWM units, serial interfaces, DMA (direct memory access) controllers, and others. Devices equipped with the peripherals required to generate and receive the external signals are often called digital signal controllers (DSC).

The algorithms and relations of digital current controllers are executed once within each sampling period T_S. The software tasks are organized within one high-priority routine called the *interrupt routine*. In Figs. 6.1 and 6.2, the execution of interrupt routines is denoted by *EXE*. Each interrupt is triggered by a dedicated, programmable event called the *interrupt tick*. In Figs. 6.1 and 6.2, the interrupt ticks are spaced by one sampling time T_S. Upon each tick, the processor interrupts the low-priority background activity and starts executing the interrupt routine. When the interrupt routine tasks are completed, the processor proceeds with the low-priority routines that were interrupted.

The tasks of the current control interrupt include the acquisition and processing of the feedback samples, the associated coordinate transformations, discrimination of the current error, calculation of the voltage references, and calculations of the pulse width modulation. Considerations and developments within this section are related to digital current controllers which use one-PWM-period feedback averaging. The overall transport delay is calculated for the conventional scheduling. Based upon that, a new scheduling scheme is considered, which reduces the transport delay.

6.1.1   Conventional Scheduling

Conventional scheduling of the current control tasks is given in Fig. 6.1. At the bottom, the voltage pulses commutate at instants denoted by ①, ②, ③ and ④. These instants are determined by the intersection of the triangular PWM carrier (the top trace in the figure) and the modulation indices u_{n-2}, u_{n-1}, u_n, and u_{n+1}, which represent the voltage references.

The interrupt ticks (*triggers*) in Fig. 6.1 coincide with the zero-count and the period-count instants of the triangular PWM carrier. The zero-count event takes places when the PWM carrier reaches zero and changes the counting direction from *down* to *up*. The period-count event takes place when the PWM carrier reaches the period value and changes the counting direction from *up* to *down*. Both events trigger T_S-spaced interrupts. For convenience, the actual values of the output current at instants of the interrupt ticks are denoted by i_{n-1}, i_n, i_{n+1}, and i_{n+2}.

The values of the voltage references u_{n-2}, u_{n-1}, u_n, and u_{n+1} are calculated in interrupt routines triggered at instants $(n-2)T_S$, $(n-1)T_S$, nT_S, and $(n+1)T_S$, respectively. The interrupt triggered at $(n+1)T_S$ calculates the voltage reference u_{n+1}, which affects the commutation instant ④, thus affecting the average value of the voltage within the interval $[(n+2)T_S .. (n+3)T_S]$. The voltage reference u_{n+1} cannot be used as the modulation index within the interval $[(n+1)T_S .. (n+2)T_S]$, since the value of u_{n+1} is not available at the beginning of the interval. It gets available later, when the interrupt routine $(n+1)$ concludes the execution. By then,

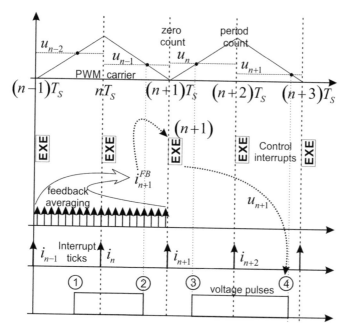

Fig. 6.1 Conventional scheduling of the current controller which relies on one-PWM-period feedback averaging

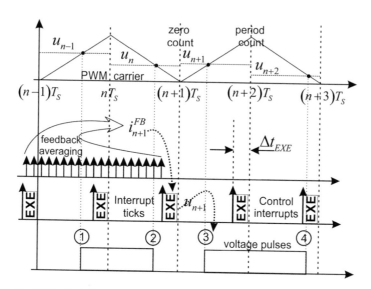

Fig. 6.2 Modified scheduling with the interrupt triggered Δt_{EXE} before the zero-count and the period-count of the PWM carrier

desired instants for the commutation ③ may have already passed. Therefore, the value u_{n+1} gets implemented one sampling period later, thus introducing the transport delay into the closed loop.

The interrupt routines receive the feedback signal i^{FB}, and they use this signal to calculate the voltage reference. The interrupt triggered at $(n+1)T_S$ operates with the feedback signal i^{FB}_{n+1}, which is obtained from the feedback averaging of N_{OV} samples acquired within the previous PWM period, namely, from $(n-1)T_S$ until $(n+1)T_S$.

By considering the fact that the feedback signals acquired on the interval $[(n-1)T_S .. (n+1)T_S]$ get processed within the interrupt $(n+1)T_S$ and affect the voltage reference u_{n+1}, which, in turn, affects the average voltage within the interval $[(n+2)T_S .. (n+3)T_S]$, the overall transport delay of the scheduling in Fig. 6.1 can be estimated to $5/2 \cdot T_S$.

6.1.2   Advanced Scheduling

With first microcontrollers that were used in digital control applications, computation delays were significant. In many cases, frequently used mathematical operations such as the multiplication were not implemented by fast hardware multipliers. In first microcontrollers, they were implemented by executing the *microcode*, a dedicated program written within the internal ROM memory within the microcontroller. In other words, multiplication required the execution of dedicated routine, taking considerable number of machine cycles. Therefore, in first digital controllers, the duration of interrupt routines (denoted by EXE in Figs. 6.1 and 6.2) could have taken more than one half of the sampling period T_S. At the same time, slow execution and large computation delays were the reason why it used to be difficult to operate with short sampling periods. Among other reasons, large computation delays of the first digital microcontrollers contributed to an extended life of analogue current controllers. With contemporary digital signal controllers, the floating-point operations complete in several nanoseconds, while complete current control interrupt routine could complete in a couple of microseconds. Thus, duration Δt_{EXE} of the interrupt is just a small fraction of the sampling period T_S. This circumstance can be used in devising advance scheduling schemes.

In Fig. 6.2, the interrupt ticks do not coincide with the zero-count and the period-count of the PWM carrier. Instead, the ticks are programmed to start the interrupts before the zero-count and the period-count events. All the interrupts are triggered Δt_{EXE} earlier. In this way, the execution of the control tasks is completed before the zero-count and the period-count events. The voltage reference u_{n+1} is calculated within the interrupt that starts at $(n+1)T_S - \Delta t_{EXE}$, and it can be loaded into the PWM peripheral before the instant $(n+1)T_S$, when the PWM carrier reaches the zero-count. Instead of being used within the interval $[(n+2)T_S .. (n+3)T_S]$, such as the case in Fig. 6.1, the reference u_{n+1} can be used earlier, within the interval $[(n+1)T_S .. (n+2)T_S]$. In this way, the transport delay is reduced by one sampling period T_S. The voltage pulses shown at the bottom of Fig. 6.2 commutate at instants denoted by ①, ②, ③ and ④, determined by the intersection of the triangular PWM carrier and the modulation indices u_{n-1}, u_n, u_{n+1}, and u_{n+2}. Compared to the schedule of Fig. 6.1, all the modulation indices get applied one sampling period earlier.

In Fig. 6.1, the feedback signal i^{FB}_{n+1} gets calculated from the current samples acquired on the interval $[(n-1)T_S \,.. \, (n+1)T_S]$. In Fig. 6.2, the samples are acquired within the interval which is time-shifted by Δt_{EXE}. Namely, the feedback signal i^{FB}_{n+1} of Fig. 6.2 gets calculated from the current samples acquired on the interval $[(n-1)T_S - \Delta t_{EXE}) \,.. \, (n+1)T_S - \Delta t_{EXE}]$. Compared to the scheme in Fig. 6.1, the feedback signal is delayed by Δt_{EXE}. Taking into account that the sampling period T_S is larger than the computation delay Δt_{EXE} by an order of magnitude ($T_S >> \Delta t_{EXE}$), delay Δt_{EXE} can be neglected. This claim has been supported by the experimental evidence provided later on within this chapter.

In Fig. 6.2, the feedback signal acquired on the interval $[(n-1)T_S - \Delta t_{EXE}) \,.. \, (n+1)T_S - \Delta t_{EXE}]$ gets used to calculate the voltage reference u_{n+1}, which determines the average output voltage on the interval $[(n+1)T_S \,.. \, (n+2)T_S]$. Neglecting Δt_{EXE}, the average transport delay reduces to $3/2 \cdot T_S$.

The new scheduling of the control tasks changes the relevant pulse transfer function and affects the design of the current controller.

6.2   Pulse Transfer Function with Advanced Scheduling

Modified scheduling scheme of Fig. 6.2 alters the pulse transfer function of the load, reducing the equivalent transport delay from $5/2T_S$ to $3/2T_S$. Analytical considerations within this section derive the modified pulse transfer functions of the load, apply the IMC approach to design the closed-loop current controller, provide the corresponding parameter setting, and evaluate the outcome, the closed-loop performances, and the disturbance rejection capability.

6.2.1   Pulse Transfer Function of the Load

In Fig. 6.2, the voltage reference u^*_{n+1} gets applied within the interval $[(n+1)T_S \,.. \, (n+2)T_S]$, where it affects the voltage pulse width and the average value of the output voltage. Similarly, the reference u^*_n gets applied within the interval $[nT_S \,.. \, (n+1)T_S]$. With the load modeled as the series connection of the back-electromotive force, the series resistance, and the series inductance, the differential equation that describes the change of the output current is

$$Ri^s(t) + L\frac{di^s(t)}{dt} = u^s(t) - e^s(t), \tag{6.1}$$

where all the variables are represented in $\alpha - \beta$ frame. Assuming that e^s_n stands for the average value of the electromotive force over the interval $[nT_S \mathinner{\ldotp\ldotp} (n+1)T_S]$, and introducing $\beta = RT_S/L$, the difference equation expressing the change of the stator current within the interval $[nT_S \mathinner{\ldotp\ldotp} (n+1)T_S]$ becomes

$$i^s_{n+1} = i^s_n e^{-\beta} + \frac{1 - e^{-\beta}}{R}\left(u^s_n - e^s_n\right) \tag{6.2}$$

Complex vectors in stationary and synchronous frames are related by

$$i^s_{n+1} = i^e_{n+1} \cdot e^{j\theta_{n+1}}, \quad i^s_n = i^e_n \cdot e^{j\theta_n}, \quad u^s_n = u^e_n \cdot e^{j\theta_n}, \tag{6.3}$$

where θ_n represents the angle between the d-axis and the α-axis at instant nT_S. Assuming that the speed changes within one sampling period are negligible, $\theta_{n+1} = \theta_n + \omega_e T_S$ and $\theta_n = \theta_{n-1} + \omega_e T_S$. The average value of the electromotive force within the interval $[nT_S \mathinner{\ldotp\ldotp} (n+1)T_S]$ becomes

$$e^e_n = \frac{1}{T}\int\limits_{nT_S}^{(n+1)T_S} e^e(t) \cdot dt \approx e^s_n \cdot e^{-j(\theta_{n+1}+\theta_n)/2}. \tag{6.4}$$

By introducing the results (6.3) and (6.4) into (6.2) and dividing the result by exp $(j\theta_n)$, the difference equation becomes

$$i^e_{n+1} \cdot e^{j\omega_e T_S} = i^e_n \cdot e^{-\beta} + \frac{1 - e^{-\beta}}{R}\left(u^e_n - e^e_n \cdot e^{j\omega_e T_S/2}\right). \tag{6.5}$$

With $\beta = RT_S/L < <1/100$, the value $(1 - e^{-\beta})/R$ can be approximated T_S/L,

$$i^e_{n+1} \cdot e^{j\omega_e T_S} = i^e_n \cdot e^{-\beta} + \frac{T_S}{L}\left(u^e_n - e^e_n \cdot e^{j\omega_e T_S/2}\right). \tag{6.6}$$

Applying the z-transform and thus converting the difference Eq. (6.6) into an algebraic equation in z-domain,

$$z \cdot i^e(z) \cdot e^{j\omega_e T_S} = i^e(z) \cdot e^{-\beta} + \frac{T_S}{L}\left(u^e(z) - e^e(z) \cdot e^{j\omega_e T_S/2}\right). \tag{6.7}$$

where the variables are the complex images $i^e(z)$, $u^e(z)$, and $e^e(z)$. The load transfer function in z-domain is obtained by dividing the complex images $i^e(z)$ and $u^e(z)$ in conditions where the disturbance $e^e(z)$ is equal to zero:

$$W_L(z) = \frac{i^e(z)}{u^e(z)} \bigg|_{e^e=0} = \frac{T_S}{L\, z \cdot e^{j\omega_e T_S} - e^{-\beta}}. \tag{6.8}$$

When both the voltage $u^e(z)$ and the back-electromotive force $e^e(z)$ are taken into consideration, the complex image of the output current in $d-q$ frame is

$$i^e(z) = W_L(z) \cdot u^e(z) - e^{\frac{j\omega_e T_S}{2}} W_L(z) \cdot e^e(z) \tag{6.9}$$

The expression (6.9) can be used for the controller design in $d-q$ frame, based on the internal model principles.

6.2.2   Design of the Controller Structure

The internal model control requires the inversion of the pulse transfer function of the load. Attempted controller is designed by multiplying the inverted pulse transfer function of the load by an integrator with an adjustable gain, $\alpha_D \cdot z/(z-1)$. The pulse transfer function of the attempted controller is

$$W_{REGX}(z) = \frac{1}{W_L(z)} \cdot \frac{\alpha_D z}{z-1} = \frac{\alpha_D L}{T_S} \cdot \frac{z \cdot \left(z \cdot e^{j\omega_e T_S} - e^{-\beta}\right)}{z-1}. \tag{6.10}$$

The pulse transfer function $W_{REGX}(z)$ implies prediction. Therefore, it cannot be implemented. Instead, the current controller is designed by dividing $W_{REGX}(z)$ by z,

$$W_{REG}(z) = \frac{W_{REGX}(z)}{z} = \frac{\alpha_D L}{T_S} \cdot \frac{z \cdot e^{j\omega_e T_S} - e^{-\beta}}{z-1}. \tag{6.11}$$

The $d-q$ frame current controller with W_{REG} given in (6.11) is represented by the block diagram in Fig. 6.3. The pulse transfer function of the load is given in (6.8), while the pulse transfer function of the feedback path is explained in Chap. 3 and given by

$$W_{FB}(z) = \frac{z^2 + 2z + 1}{4z^2}. \tag{6.12}$$

6.2.3   Closed-Loop and Disturbance Transfer Functions

The current controller designed in (6.11) is obtained by inverting the load dynamics W_L, multiplying the outcome by an integrator with an adjustable gain $(\alpha_D \cdot z/(z-1))$, and dividing the result by z. Therefore, the product $W_{REG}(z) \cdot W_L(z)$ becomes

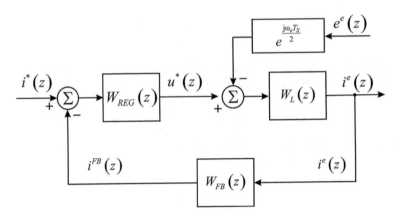

Fig. 6.3 Block diagram of the current controller in d-q frame

$$W_{\text{REG\_L}}(z) = W_{\text{REG}}(z)W_{\text{L}}(z) = \frac{\alpha_D}{z-1}. \tag{6.13}$$

From Fig. 6.3, the closed-loop transfer function is equal to

$$W_{\text{CL}}(z) = \frac{i^e(z)}{i^*(z)}\bigg|_{e^e=0} = \frac{W_{\text{REG\_L}}(z)}{1 + W_{\text{REG\_L}}(z) \cdot W_{\text{FB}}(z)}. \tag{6.14}$$

Introducing the expressions (6.12) and (6.13) into (6.14), the closed-loop pulse transfer function becomes

$$W_{\text{CL}} = \frac{i^e(z)}{i^*(z)}\bigg|_{e^e=0} = \frac{4\alpha_D z^2}{4z^3 + (\alpha_D - 4)z^2 + 2\alpha_D z + \alpha_D} = \frac{4\alpha_D z^2}{f_{\text{CP}}(z)}, \tag{6.15}$$

where $f_{\text{CP}}(z)$ is characteristic polynomial. In a like manner, it is possible to obtain the disturbance pulse transfer function, which describes the impact of the voltage disturbances on the output current. This transfer function is obtained as

$$Y^e(z) = \frac{i^e(z)}{-e^e(z)}\bigg|_{i^*(z)=0}. \tag{6.16}$$

From the block diagram in Fig. 6.3,

$$Y^e(z) = \frac{e^{\text{j}\frac{1}{2}\omega_e T_S} W_{\text{L}}(z)}{1 + W_{\text{L}}(z) \cdot W_{\text{REG}}(z) \cdot W_{\text{FB}}(z)} = \frac{e^{\text{j}\frac{1}{2}\omega_e T_S} W_{\text{L}}(z)}{1 + W_{\text{REG\_L}}(z) \cdot W_{\text{FB}}(z)}. \tag{6.17}$$

By introducing the expressions for $W_{\text{FB}}(z)$, $W_{\text{REG\_L}}(z)$, and $W_{\text{L}}(z)$, the pulse transfer function Y^e becomes

$$Y^e(z) = \frac{i^e(z)}{-e^e(z)}\bigg|_{i^*=0} = \frac{T_S}{L}\frac{4 \cdot z^2(z-1) \cdot e^{\frac{j\omega_e T_S}{2}}}{z \cdot e^{j\omega_e T_S} - e^{-\beta}} \cdot \frac{1}{f_{CP}(z)}, \tag{6.18}$$

where

$$f_{CP}(z) = 4z^3 + (\alpha_D - 4)z^2 + 2\alpha_D z + \alpha_D \tag{6.19}$$

is the characteristic polynomial.

6.2.4 Parameter Setting and the Closed-Loop Performance

Decoupling current controller of (6.11) has only one gain, α_D. The closed-loop transfer function $W_{CL}(z)$ is given in (6.15). The key performance indices include the vector margin, the closed-loop bandwidth frequencies $f_{BW}(45°)$ and $f_{BW}(-3dB)$, as well as the overshoot of the input step response. These performance indices are given in Table 6.1 for the closed-loop pulse transfer function of (6.15) and for the gain α_D that changes from 0.2 up to 0.45. Comparable closed-loop transfer function (5.25) is obtained with conventional scheduling of Fig. 6.1, while the corresponding results are given in Table 5.2.

The schedule of Fig. 6.1 in conjunction with the closed-loop transfer function $W_{CL}(z)$ of (5.25) and with the gain $\alpha_D = 0.25$ reaches the closed-loop bandwidth $f_{BW}(-3dB) = 0.0939/T_S$ with an overshoot of 14% and with the vector margin of 0.5664. At the same time, the current controller $W_{REG}(z)$ of (6.11) with the schedule of Fig. 6.2 and with the closed-loop transfer function of (6.15) has a comparable overshoot of 12% with the closed-loop bandwidth $f_{BW}(-3\ dB) = 0.1447/T_S$, with the vector margin of 0.608, and with the gain of $\alpha_D = 0.4$. Thus, the schedule of Fig. 6.2 allows a wider range of applicable gains; it increases the closed-loop bandwidth, reduces the overshoot, and increases the vector margin.

Simulation traces of the input step response obtained with the gain α_D ranging from 0.2 up to 0.45 are given in Fig. 6.4. From the comparison of the traces shown in Fig. 5.7 and the ones of Fig. 6.4, one concludes that the new scheduling of Fig. 6.2 increases the range of applicable gains.

It is also of interest to explore the impact of the new scheduling scheme of Fig. 6.2 on the disturbance rejection capability. The pulse transfer function $Y^e(z)$, given in (6.18), defines the response of the output current $i^e(z)$ to the changes of the voltage disturbances, represented in Fig. 6.3 as the back-electromotive force $e^e(z)$.

In operating regime where the current reference i^* is equal to zero, any output current caused by the voltage disturbance is, as a matter of fact, the current error. Assuming that the voltage disturbance is a Heaviside step, namely, that $e^e(z) = -E_d/(1 - z^{-1})$, the integral of the output-current error becomes

Table 6.1 Closed-loop performance obtained with the decoupling controller of (6.11) and with the closed-loop pulse transfer function of (6.15)

Gain α_D	VM	$f_{BW}(45°)$	$f_{BW}(-3\ dB)$	Overshoot
0.20	0.7820	$0.0341/T_S$	$0.0497/T_S$	0
0.25	0.7359	$0.0429/T_S$	$0.0726/T_S$	0.0008
0.30	0.6917	$0.0516/T_S$	$0.0987/T_S$	0.0246
0.35	0.6492	$0.0600/T_S$	$0.1234/T_S$	0.0676
0.40	0.6081	$0.0683/T_S$	$0.1447/T_S$	1.1236
0.45	0.5683	$0.0762/T_S$	$0.1627/T_S$	1.1868

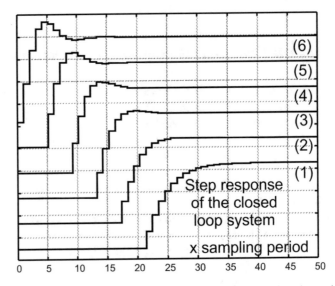

Fig. 6.4 Input step response from the pulse transfer function of (6.15) for the gain equal to $\alpha_D = 0.2$ (1), $\alpha_D = 0.25$ (2), $\alpha_D = 0.3$ (3), $\alpha_D = 0.35$ (4), $\alpha_D = 0.4$ (5), and $\alpha_D = 0.45$ (6)

$$I(\infty) = \lim_{z \to 1} \left[(z-1) \cdot \frac{z}{z-1} \cdot \frac{E_d \cdot z}{z-1} \cdot Y^e(z) \right]. \tag{6.20}$$

The result of (6.20) depends on the output frequency ω_e. Assuming that the output frequency is considerably lower than the sampling frequency, the factors such as exp $(j\omega_e T_S)$ are close to 1. Introducing $Y^e(z)$ from (6.18),

$$I(\infty) = \lim_{z \to 1} \left[E_d \frac{T_S}{L} \frac{4 \cdot z^4 \cdot e^{\frac{j\omega_e T_S}{2}}}{z \cdot e^{j\omega_e T_S} - e^{-\beta}} \cdot \frac{1}{f_{CP}(z)} \right]$$

$$\approx \lim_{z \to 1} \left[E_d \frac{T_S}{L} \frac{4 \cdot z^4}{z - e^{-\beta}} \cdot \frac{1}{f_{CP}(z)} \right]. \tag{6.21}$$

For $z = 1$, $f_{CP}(z) = 4\alpha_D$, while $z - \exp(-\beta) \approx RT_S/L$. Therefore,

$$I(\infty) = \frac{E_d}{\alpha_D R}. \tag{6.22}$$

Since the current controller with advanced scheduling of Fig. 6.2 has an increased range of applicable gains α_D, it also provides lower values of $I(\infty)$ and better disturbance rejection capability.

6.3 Advanced Scheduling with Series Compensator

The advanced scheduling of Fig. 6.2 reduces the transport delay from $5/2T_S$ to $3/2T_S$. Reduction of the delay has beneficial effects on the closed-loop performance. According to the results in Table 6.1, the closed-loop bandwidth is increased, the overshoot of the input step response is reduced, while the vector margin and the robustness of the controller are both improved. Further improvements can be achieved by adding the series differential compensator, introduced in Sect. 5.3. The effect of delays can be reduced by introducing the differential action which improves the phase characteristic, increases the damping factor, and reduces the overshoot. The current controller discussed in Sect. 6.2 is extended by adding the series differential compensator. The consequential closed-loop performance with the series differential compensator is evaluated analytically and by means of computer simulations.

The block diagram of the current controller with series differential compensator is given in Fig. 6.5. The pulse transfer functions $W_{REG}(z)$, $W_{FB}(z)$, and $W_L(z)$ remain the same as the ones of Fig. 6.3. The only difference is the insertion of the series compensator $W_{DIF}(z)$ of differential nature.

6.3.1 Closed-Loop and Disturbance Transfer Functions

Series differential compensator of (5.30) has the pulse transfer function

$$W_{DIF}(z) = 1 + d\frac{z-1}{z} = \frac{(1+d)z - d}{z}, \tag{6.23}$$

where d is an adjustable gain. The product of the current controller W_{REG} designed in (6.11), the pulse transfer function of the load W_L, and the pulse transfer function of the differential compensator is

$$W_{REG\_L\_DIF}(z) = W_{REG}(z) \cdot W_L(z) \cdot W_{DIF}(z) = \frac{\alpha_D}{z-1} \cdot \frac{(1+d)z - d}{z}. \tag{6.24}$$

From Fig. 6.5, the closed-loop transfer function is equal to

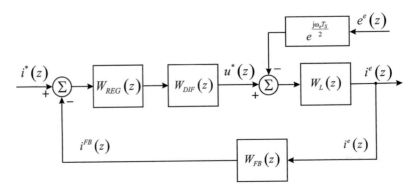

Fig. 6.5 Block diagram of the current controller in d-q frame with series differential compensator

$$W_{CL}(z) = \left.\frac{i^e(z)}{i^*(z)}\right|_{e^e=0} = \frac{W_{REG\_L\_DIF}(z)}{1 + W_{REG\_L\_DIF}(z) \cdot W_{FB}(z)}. \tag{6.25}$$

Introducing the expression (6.24) into (6.25), the closed-loop pulse transfer function becomes

$$W_{CL}(z) = \left.\frac{i^e(z)}{i^*(z)}\right|_{e^e=0} = \frac{4\alpha_D(1+d)z^3 - 4\alpha_D dz^2}{f_{CP}(z)}$$
$$= \frac{4\alpha_D(1+d)z^3 - 4\alpha_D dz^2}{4z^4 + [\alpha_D(1+d) - 4]z^3 + \alpha_D(2+d)z^2 + \alpha_D(1-d)z - \alpha_D d}. \tag{6.26}$$

where $f_{CP}(z)$ is the characteristic polynomial. Similarly, it is possible to obtain the disturbance pulse transfer function $Y^e(z)$. From the block diagram in Fig. 6.5,

$$Y^e(z) = \frac{e^{j\frac{1}{2}\omega_e T_s} W_L(z)}{1 + W_{REG\_L\_DIF}(z) \cdot W_{FB}(z)}. \tag{6.27}$$

By introducing the expressions for $W_{FB}(z)$, $W_{REG\_L\_DIF}(z)$, and $W_L(z)$, the pulse transfer function $Y^e(z)$ becomes

$$Y^e(z) = \left.\frac{i^e(z)}{-e^e(z)}\right|_{i^*=0} = \frac{T_s}{L} \frac{4 \cdot z^3(z-1) \cdot e^{\frac{j\omega_e T_s}{2}}}{z \cdot e^{j\omega_e T_s} - e^{-\beta}} \cdot \frac{1}{f_{CP}(z)}, \tag{6.28}$$

where

$$f_{CP}(z) = 4z^4 + [\alpha_D(1+d) - 4]z^3 + \alpha_D(2+d)z^2 + \alpha_D(1-d)z - \alpha_D d \tag{6.29}$$

is the characteristic polynomial.

6.3.2   Parameter Setting and the Closed-Loop Performance

The current controller enhanced with the series differential compensator has two gains, α_D and d. The closed-loop transfer function $W_{CL}(z)$ is given in (6.26). For $\alpha_D = 0.4$ and for the differential gain d ranging from 0 up to 1, the key performance indices are listed in Table 6.2. These indices include the vector margin, the closed-loop bandwidth frequencies $f_{BW}(45°)$ and $f_{BW}(-3\text{dB})$, as well as the overshoot of the input step response. The Matlab script which is used to calculate the relevant performance indices is listed in Table 6.3.

According to the results in Table 6.2 and according to the comparison of these results to the ones obtained in Table 6.1, one concludes that the introduction of the series differential compensator provides a further increase of the closed-loop performances. Namely, it is possible to reach the closed-loop bandwidth that exceeds 22% of the sampling frequency while keeping the vector margin beyond 0.6 and maintaining the overshoot below 3%. Thus, with the IMC-based controller, with the advanced scheduling of the control tasks, and with the series differential compensator, the closed-loop bandwidth of digital current controllers applied to the grid-side inverters with $f_{PWM} = 10$ kHz, the closed-loop bandwidth exceeds $0.22 \cdot f_S = 0.22 \cdot 20000 = 4.4 \ kHz$. With a closed-loop bandwidth of that magnitude, the current controller is more capable of suppressing the line harmonics and other disturbances.

In Table 6.2, the differential gain d has beneficial effects until reaching 0.6. Larger values reduce the vector margin, and they start increasing the overshoot, which increases to 4% for $d = 1$. Therefore, it is of interest to observe the waveforms of the input step responses, obtained with the gains that are considered in Table 6.2. Simulation traces of the input step response obtained with the gain $\alpha_D = 0.4$ and with the gain d ranging from 0 up to 1 are given in Fig. 6.6. For $d = 0.6$ and $d > 0.6$,

In addition to considering the input step response, it is also of interest to analyze the disturbance rejection capability. The relevant pulse transfer function $Y^e(z)$ is given in (6.28). Assuming that the voltage disturbance is a Heaviside step, namely, that $e^e(z) = -E_d/(1 - z^{-1})$, the integral of the output-current error is given in (6.20). Assuming that the output frequency ω_e is considerably lower than the sampling frequency, the factors such as exp $(j\omega_e T_S)$ are close to 1. Introducing $Y^e(z)$ from (6.20) and assuming that $1 - \exp(-\beta) \approx RT_S/L$, the integral of the current error caused by the voltage step E_d is defined by quite the same as the result of (6.22). The new scheduling and the differential compensator provide an increased range of applicable gains α_D, thus resulting in lower values of $I(\infty)$ and an improved disturbance rejection capability.

$$I(\infty) = \frac{E_d}{\alpha_D R}, \tag{6.30}$$

Table 6.2 Closed-loop performance obtained with the decoupling controller of (6.11), with series differential compensator, and with the closed-loop pulse transfer function of (6.26)

Gain α_D	Gain d	VM	$f_{BW}(45°)$	$f_{BW}(-3\ dB)$	Overshoot
0.4	0	0.6081	$0.0683/T_S$	$0.1447/T_S$	0.1236
0.4	0.2	0.6328	$0.0738/T_S$	$0.1578/T_S$	0.0766
0.4	0.4	0.6399	$0.0821/T_S$	$0.1831/T_S$	0.0353
0.4	0.6	0.6296	$0.0945/T_S$	$0.2214/T_S$	0.0212
0.4	0.8	0.6075	$0.1115/T_S$	$0.2610/T_S$	0.0105
0.4	1	0.5787	$0.1305/T_S$	$0.2989/T_S$	0.0400

Table 6.3 Matlab code which calculates the vector margin, the closed-loop bandwidth, and the overshoot in the step response of $W_{CL}(z)$ of (6.26)

```
a = 0.40;                            % Set the parameter αD
d = 0.0;                             % Set the differential gain

num = [4*a*(1+d)-4*a*d 0 0];         % Numerator of WCL of (6.26)
den = [4(a+a*d-4)(2*a+d*a)(a-a*d)-a*d]; % Denominator of WCL of (6.26)
iout = dstep(num,den,50);            % Input step response

for jj=1:1000, W(jj)=jj*50; end;     % Generating the vector W [0 .. Wmax]
z = exp(1i*W*T);                     % Generating the vector z
WPP=a./(z-1).*(1 + 2./z + 1./z./z)/4; % Calculating the open loop gain for
WPP = WPP.*((1+d).*z - d)./z;        % W sweeping from 0 to Wmax
VM = min(abs(1+WPP));                % Calculating the vector margin
Preb = max(iout);                    % Calculating the overshoot

[Mag,Pha,W] = dbode(num,den,0.00005,W); % Getting the amplitude and
                                     % frequency characteristic
f45 = W(min(find(Pha< -45)))/2/pi;   % from Bode-plot. Getting
fbw = W(min(find(Mag<0.707)))/2/pi;  % the bandwidth

disp('a  d  VM  f45  f3db  preb ')   % Printing the outputs

[a  d  VM  f45*0.00005  fbw*0.00005 Preb]
disp('a d VM f45 f3db preb')
[a d min(abs(1+WPP))  f45*T fbw*T Preb]
```

6.4   Experimental Results

This chapter introduced an advanced scheduling of the digital current control tasks. The new scheduling reduces the transport delays, and it contributes to significant increase of the closed-loop bandwidth. The application of the new scheduling, the analytical considerations, and the parameter setting are all based upon the assumption that the computation delay Δt_{EXE} which is an order of magnitude shorter than the sampling period T_S does not have any meaningful effect. In addition to the analytical considerations and computer simulations, it is also of interest to verify this fundamental assumption experimentally.

The new scheduling in conjunction with the series differential compensator provides with the possibility of reaching very high closed-loop bandwidth that exceeds 20% of the sampling frequency. It is of interest to verify such analytical findings

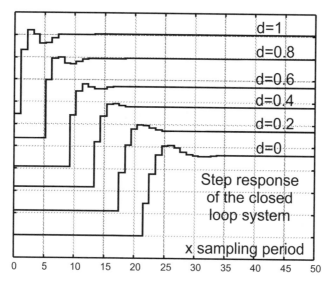

Fig. 6.6 Input step response from the pulse transfer function of (6.26) for the gain of $\alpha_D = 0.4$ and for the differential gain ranging from 0 up to 1

experimentally. The experimental setup used for the test and the relevant parameters are described in Sect. 5.4.1. The experiments were performed by introducing step changes into the reference currents in $d-q$ frame. The waveforms represent the output currents in $d-q$ frame, stored into the RAM memory of the DSP controller. Thus, all the signals are obtained at the output of the feedback chain, and they correspond to $i^{FB}(z) = W_{FB}(z) \cdot i^e(z)$ The sampling time T_S is 50 µs, and the switching frequency is $f_{PWM} = 10$ kHz.

6.4.1   The Impact of the Computation Delay

The purpose of experimental verifications performed in this section is to investigate the impact of the computation delay Δt_{EXE} on the dynamic performances of the controllers with the new, advanced schedule and to establish viable limit for the ratio $\Delta t_{EXE}/T_S$. The crucial hypothesis introduced with the new scheduling is that the delay Δt_{EXE} of the feedback signal $i^{FB}{}_{n+1}$ in Fig. 6.2 does not have any meaningful effect on the dynamic behavior. This hypothesis is experimentally tested and verified by performing the step response test and varying Δt_{EXE} over a wide range. With conventional controllers such as the DSC TMS320F28335 and with the use of the optimizing compiler, the control and protection tasks complete in less than 4 µs. For the purposes of this tests, some control and protection routines unrelated to the current control are removed in order to reduce the computation delay, which enabled the operation with Δt_{EXE} as low as 2.4 µs. During the test runs, Δt_{EXE} is varied from 2.4 µs up to 12 µs. The sampling time T_S is set to 50 µs.

The traces in Figs. 6.7 and 6.8 represent the step response of q-axis current. The traces in Fig. 6.7 correspond to the current controller developed in Sect. 6.2, which uses the advance scheduling and operates without the series differential compensator. The traces in Fig. 6.8 are obtained with the controller developed in Sect. 6.3, which uses the advance scheduling and also the series differential compensator.

In both figures, the trace (1) is obtained from the computer simulation. The simulation was performed assuming that $\Delta t_{EXE} = 0$. Thus, the trace (1) serves as the reference for the comparison. The waveform (2) is obtained experimentally, with the shortest possible $\Delta t_{EXE} = 2.5$ µs. The remaining traces were obtained by increasing Δt_{EXE} deliberately, in order to identify the ratio $\Delta t_{EXE}/T_S$ where the computation delay starts to affect the input step response of the closed-loop system. The trace (3) is obtained with $\Delta t_{EXE} = 4$ µs, the trace (4) with $\Delta t_{EXE} = 8$ µs, and the trace (5) with $\Delta t_{EXE} = 12$ µs.

In both Figs. 6.7 and 6.8, the traces (2) and (3) are practically unaffected by the computation delay Δt_{EXE}. The step response is noticeably changed in trace 5 of Fig. 6.7, where Δt_{EXE} exceeds $0.2 \cdot T_S$. Similarly, the response significantly deteriorates in Fig. 6.8 for traces 4 and 5, where Δt_{EXE} equals 8µs and 12µs, respectively. With only a minor difference between reference trace 1 and experimental traces 2 and 3, one concludes that the time shift $\Delta t_{EXE} = 0.08 \cdot T_S$ does not have any significant impact on the closed-loop dynamics. This corroborates the assumption introduced in this chapter.

6.4.2 Input Step Response

Input step response of the q-axis current and the contemporary waveform of the d-axis current are obtained in regime where the modulation indices and the output voltage reach 90% of their maximum value. The three-phase inverter supplies a brushless dc motor. The traces (1) and (2) were obtained with the current controller that uses the advance scheduling, but it does not use the series differential compensator. The traces (3) and (4) where obtained by adding the series differential compensator. In both cases, the step response is quick, it does not have an overshoot, while the transients in q-axis do not disturb the current in d-axis (Fig. 6.9).

The traces of the q-axis current are enlarged and plotted again in Fig. 6.10, along with the simulation traces, added for the purposes of the comparison. The traces (1) and (2) correspond to the current controller with the advanced scheduling and without a series differential compensator. The traces (3) and (4) are obtained with the current controller that includes the series differential compensator. Simulation traces (2) and (4) provide the waveform of the feedback signal $i^{FB}(z)$, obtained at the output of the block $W_{FB}(z)$ in Figs. 6.3 and 6.5. The experimental traces (1) and (3) are obtained from the experimental setup, by logging the feedback signal i^{FB} into the RAM memory of the DSP controller. Dynamic response of the simulation traces and

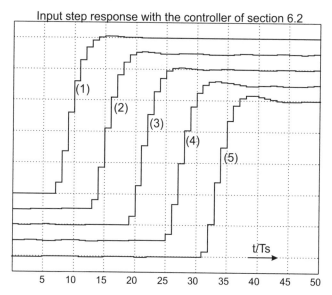

Fig. 6.7 Input step response of the *q*-axis current obtained with the current controller of Sect. 6.2, which uses the advanced scheduling of the control tasks, and it does not use any series compensator. The trace 1 is obtained by computer simulation, and it serves for the reference. For the experimental traces 2–5, the time shift Δt_{EXE} of the execution of the control interrupt is set to 2.4 μs, 4 μs, 8 μs, and 12 μs, respectively

the experimental traces is rather similar. In steady state, the simulation traces do not have any disturbances, while the experimental traces have a certain amount of jitter caused by the noise and the finite resolution.

It is of interest to verify the operation of the current controller at very high output frequencies. In Fig. 6.11, the waveforms present the step response of the current controller at the output frequency of $f_e = 0.1/T_S$. The three-phase inverter is loaded with the three star-connected inductances. The waveforms in Fig. 6.11 present two pairs of traces, $i_d(t)$ and $i_q(t)$, for two different controller structures. The two uppermost traces are obtained with the current controller which uses the advanced scheduling without the differential compensator. The bottom two traces correspond to the current controller with the series differential compensator. The waveforms demonstrate the capability of the devised controllers to provide the decoupled operation even at very high output frequencies.

6.4.3   Robustness Against the Parameter Changes

In current controllers where the closed-loop bandwidth is increased on account of an increased gain, it is of interest to verify the robustness against the parameter changes, that is, the capability of the controller to provide a stable, well-damped response

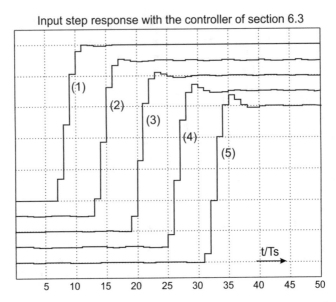

Fig. 6.8 Input step response of the q-axis current obtained with the current controller of Sect. 6.3, which uses the advanced scheduling and also the series differential compensator. The trace 1 is obtained by computer simulation, and it serves for the reference. For the experimental traces 2–5, the time shift Δt_{EXE} of the execution of the control interrupt is set to 2.4 µs, 4 µs, 8 µs, and 12 µs, respectively

even in cases where the system parameters change over a wide range. Such a system parameter is often the load inductance. It can change due to magnetic saturation, as well as due to parasitic effects. Analytical proof of the system robustness is obtained by calculating the vector margin. In Fig. 6.12, the robustness is tested by observing the experimental traces of the q-axis current in conditions where the system parameters exhibit significant changes. The traces represent the step reponse of the q-axis current and the contemporary waveform of the d-axis current, obtained in regime where the modulation indices and the input step response in the presence of parameter change. The traces are obtained with the current controller which uses the advanced scheduling without the differential compensator. The ratio between the actual load inductance L and the parameter L_{nom} is changed from 0.6 up to 1.5. The parameter L_{nom} is used within the algorithm to calculate the gain $(\alpha_D L/T_S)$. Similar traces are plotted in Fig. 6.13, this time with the current controller developed in Sect. 6.3, which uses the advanced scheduling and the series differential compensator. In both cases, the response to the input step is altered by the parameter change, but the system maintains a stable, well-damped response.

Question (6.1) For the discrete-time current controller with advanced scheduling of the control tasks, with the closed-loop transfer function given in (6.15), determine the maximum gain α_D which results in conjugate complex closed-loop pole pair in s-domain with the damping of $\xi > 0.7$. Instead of an analysis, it is advisable to use the

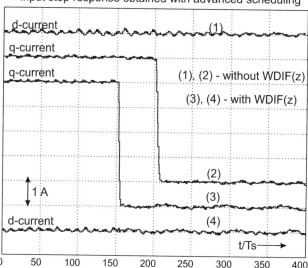

Fig. 6.9 Input step response of the q-axis current and the contemporary waveform of the d-axis current obtained in regime where the modulation indices and the output voltage reach 90% of their maximum value. The three-phase inverter supplies a brushless dc motor. The traces (1) and (2) were obtained with the current controller that uses the advance scheduling, but it does not use the series differential compensator. The traces (3) and (4) where obtained by adding the series differential compensator

computer math tools and, by increasing the gain gradually, find the approximate value of α_D which gives $\xi = 0.7$.

Answer (6.1) The closed-loop transfer function is given in

$$W_{CL}(z) = \frac{4\alpha_D z^2}{4z^3 + (\alpha_D - 4)z^2 + 2\alpha_D z + \alpha_D}$$

The characteristic polynomial has three poles. Relation between the poles in s-domain and the poles in z-domain is.

$$z_1 = e^{s_1 T}, \quad z_2 = e^{s_2 T}$$

The following sequence of Matlab commands provides the information on the z-domain poles:

```
>> fsampling = 10000;
>> Gain = 0.3;
>> zz = roots([4 (Gain-4) 2*Gain Gain]);
>> z1 = zz(1); z2 = zz(2); z3 = zz(3);
```

Input step response obtained with advanced scheduling

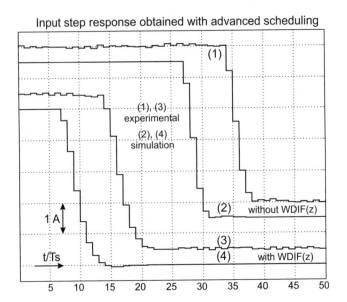

Fig. 6.10 Comparison of the experimental traces and the simulated traces. The traces represent the input step response of the q-axis current. The experimental traces are obtained by enlarging the traces (2) and (3) of Fig. 6.9

Input step response obtained with advanced scheduling

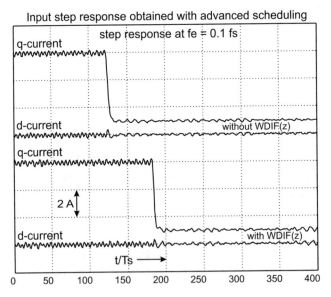

Fig. 6.11 Step response of the current controller at very high output frequency of $f_e = 0.1 \cdot f_S$. The three-phase inverter is loaded with the three star-connected inductances. The two uppermost traces are obtained with the current controller which uses the advanced scheduling without the differential compensator. The bottom two traces correspond to the current controller with the series differential compensator

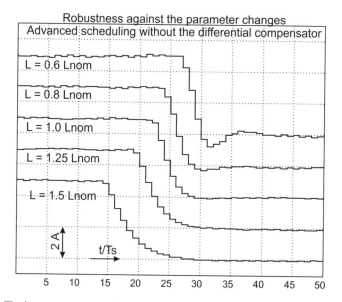

Fig. 6.12 The input step response in the presence of parameter changes. The traces are obtained with the current controller which uses the advanced scheduling without the differential compensator. The traces represent the step response of the q-axis current. The ratio between the actual load inductance L and the parameter L_{nom} is changed. The parameter L_{nom} is used within the algorithm to calculate the gain ($\alpha_D L/T_S$)

By altering the gains and studying the closed-loop poles, it is concluded that one of the poles resides on the negative part of the real axis in z-plane, on the interval $[-1 .. 0]$. For the gains larger than 0.2268, the other two poles are conjugate complex poles. Their damping factor is the ratio between the real part of the s-domain equivalent and the absolute value of the s-domain equivalent. The maximum gain that maintains the desired damping factor of conjugate complex poles is found from the following script:

```
>> fsampling = 10000;
>> Damping = 1; Gain = 0.1; % Initial values
>> while Damping > 0.7;
>>    Gain = Gain + 0.00001;
>>    zz = roots([4 (Gain-4) 2*Gain Gain]);
>>    zx = zz(1); if real(zx) < 1, zx = zz(2); end;
>>    Spole = 2*pi*fsampling*log(zx);
>>    Damping = -real(Spole)/abs(Spole);
>> end;
```

The above script results in Gain $= \alpha_D = 0.3298$.

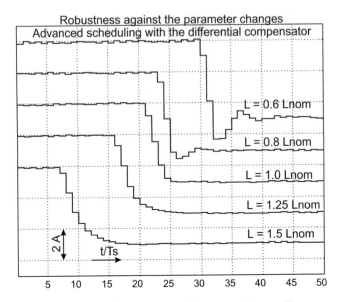

Fig. 6.13 The input step response in the presence of parameter changes. The traces are obtained with the current controller which uses the advanced scheduling with the differential compensator. The traces represent the step response of the q-axis current. The ratio between the actual load inductance L and the parameter L_{nom} is changed. The parameter L_{nom} is used within the algorithm to calculate the gain $(\alpha_D L/T_S)$

Question (6.2) Considering the system of Question 6.1, determine the vector margin for the given gain.

Answer (6.2) The vector margin is obtained from the open-loop transfer function of the system given in Fig. 6.3.

$$W_{\text{OLG}}(z) = W_{\text{REG}}(z)W_{\text{L}}(z)W_{\text{FB}}(z)$$
$$= \frac{\alpha_D}{z-1} \cdot \frac{z^2 + 2 \cdot z + 1}{4 \cdot z^2}$$

The vector margin can be found by running the script given below. The vector margin found is VM = 0.67, which is in good agreement with the results given in Table 6.1.

```
>> Gain = 0.3;
>> fsampling = 10000;
>> VM = 1;
>> for j = 1:fsampling/2,
>>     Wrads = 1i*j*2*pi;
>>     z=exp(Wrads/fsampling);
>>     WOLG = Gain*(z*z+2*z+1)/z/z/(z-1)/4;
>>     WW = abs(1+WOLG);
```

```
>>    if WW < VM, VM = WW; end;
>> end
```

Question (6.3) Considering the system of Questions 6.1 and 6.2 and using the computer math tools, determine the maximum gain αD which results in the step response with an overshoot lower than 25%. Determine the vector margin for this gain.

Answer (6.3): Analytical solution of the problem is quite involved. Therefore, it is of interest to use the math tools (Matlab) in order to find the solution. The following sequence of commands provides the information on the overshoot that is obtain for the given gain.

```
>> Gain = 0.25;
>> Response = dstep(4*Gain, [4 (Gain-4) 2*Gain Gain]);
>> Overshoot = max(Response) -1;
```

In order to find the largest gain with an overshoot lower than 25%, it is possible to organize the search where the gain is gradually increased:

```
>> Overshoot = 0; Gain = 0.2; % Initial values
>> while Overshoot < 0.25;
>>    Gain = Gain + 0.00005;
>>    Overshoot = max(dstep(4*Gain, [4 (Gain-4) 2*Gain Gain])-1);
>> end;
```

The above script results in Gain $= \alpha_D = 0.5$. At this point, it is necessary to calculate the vector margin (VM). The vector margin can be found by running the script given in Answer to Question 6.2. The vector margin found is VM $= 0.53$, which is in good agreement with the results given in Table 6.1.

Question (6.4) For the discrete-time current controller with advanced scheduling of the control tasks, with series differential compensator having $d = 0.7$, with the block diagram shown in Fig. 6.5, and with the closed-loop transfer function given in (6.26), determine the maximum gain α_D which results in conjugate complex closed-loop pole pair in s-domain with the damping of $\xi > 0.4$. Instead of an analysis, it is advisable to use the computer math tools and, by increasing the gain gradually, find the approximate value of α_D which gives $\xi = 0.5$.

Answer (6.4) The closed-loop transfer function is given in

$$W_{\mathrm{CL}}(z) = \frac{4\alpha_D(1+d)z^3 - 4\alpha_D d z^2}{4z^4 + [\alpha_D(1+d) - 4]z^3 + \alpha_D(2+d)z^2 + \alpha_D(1-d)z - \alpha_D d}.$$

The characteristic polynomial has four poles. The following sequence of Matlab commands provides the information on the z-domain zeros and poles,

```
>> fsampling = 10000; Gain = 0.3; d = 0.7;
>> num = [4*Gain*(1+d) -4*Gain*d 0 0];
>> den = [4 (Gain*(1+d)-4) Gain*(2+d) Gain*(1-d) -Gain*d];
>> zzeros = roots(num);
>> ppoles = roots(den);
```

By altering the gains and studying the closed-loop poles, it is concluded that one of the poles resides on the negative part of the real axis in z-plane, on the interval $[-1$ .. 0], the other real pole is positive, while other two poles are conjugate complex poles. Their damping factor is the ratio between the real part of the s-domain equivalent and the absolute value of the s-domain equivalent. The maximum gain that maintains the desired damping factor of conjugate complex poles is found from the following script:

```
>> fsampling = 10000; Gain = 0.1; d = 0.7; Damping = 1;
>> while Damping > 0.5,
>> den = [4 (Gain*(1+d)-4) Gain*(2+d) Gain*(1-d) -Gain*d];
>> ppoles = roots(den); Gain = Gain + 0.0001;
>> Damping = min(abs(real(ppoles)./abs(ppoles)))
>> end
```

The above script results in Gain $= \alpha_D = 0.3609$.

Question (6.5) Considering the system of Question 6.4, determine the vector margin for the given gain.

Answer (6.5) The vector margin is obtained from the open-loop transfer function of the system given in Fig. 6.5.

$$W_{\text{OLG}}(z) = W_{\text{REG}}(z)W_{\text{L}}(z)W_{\text{FB}}(z)W_{\text{DIF}}(z) =$$
$$= \frac{\alpha_D}{z-1} \cdot \frac{z^2 + 2 \cdot z + 1}{4 \cdot z^2} \cdot \frac{(1+d)z - d}{z}$$

The vector margin can be found by running the script given below. The vector margin is VM $= 0.6544$.

```
>> Gain = 0.3609; d = 0.7;
>> fsampling = 10000;
>> VM = 1;
>> for j = 1:fsampling/2,
>>    Wrads = 1i*j*2*pi;
>>    z=exp(Wrads/fsampling);
>>    WOLG = Gain*(z*z+2*z+1)/z/z/(z-1)/4;
```

```
>>      WOLG = WOLG*(1+d -d/z);
>>      WW = abs(1+WOLG);
>>      if WW < VM, VM = WW; end;
>> end
```

Question (6.6) Considering the system of Questions 6.4 and 6.5, with the differential gain set to $d = 0.7$, and using the computer math tools, determine the maximum gain α_D which results in the step response with an overshoot lower than 25%.

Answer (6.6) It is of interest to use the math tools (Matlab) in order to find the solution. In order to find the largest gain with an overshoot lower than 25%, it is possible to organize the search where the gain is gradually increased:

```
>> Gain = 0.2; d = 0.7; Overshoot = 0;
>> while Overshoot < 0.25,
>> Gain = Gain + 0.0005;
>> num = [4*Gain*(1+d) -4*Gain*d 0 0];
>> den = [4 (Gain*(1+d)-4) Gain*(2+d) Gain*(1-d) -Gain*d];
>> Respo = dstep(num,den);
>> Overshoot = max(Respo)-1;
>> Gain
>> end
```

Under the circumstances, the maximum gain that keeps the overshoot below 0.25 is Gain $= 0.5415$.

Question (6.7) For the gain obtained in Question 6.6, calculate the vector margin.

Answer (6.7) The vector margin can be found by running the script given in Answer to Question 6.5. For the gain setting of Question 6.6, the vector margin is VM $= 0.4974$.

Chapter 7
Disturbance Rejection

Digital current controllers have the crucial impact on performance of grid-side converters and ac drives. The tasks of the current controller include an error-free tracking of the input reference but also the suppression of the voltage disturbance. In ac drives, the voltage disturbances are the back-electromotive forces of ac machines. In grid-side inverters, the voltage disturbances are the line voltages. The voltage disturbances are commonly suppressed by enhancing the controller with an inner active resistance feedback, as described in Sect. 4.5. In cases where the switching noise and parasitic oscillations introduce sampling errors, conventional sampling is replaced by the oversampling-based error-free feedback acquisition which derives the average of the measured currents over the past switching period. This one-PWM-period feedback averaging is introduced in Sect. 3.3. The time delay introduced into the feedback path creates difficulties in designing the current controller with the active resistance. In this chapter, the possibility of applying the active resistance feedback in systems with the error-free sampling is introduced and discussed. The analysis is focused on studying the impact of transport delays, introduced by the feedback averaging on the range of the applicable active resistance gains. The internal model principle is applied in order to get the modified current controller where the active resistance gain does not affect the input step response. Disturbance rejection capability is tested analytically, by computer simulation and experimentally.

7.1 Active Resistance Feedback

The active resistance feedback is a local proportional control action described in Sect. 4.5 and illustrated in Fig. 7.1. The current feedback, whether obtained from the synchronous center-pulse sampling or the one-PWM-period averaging, gets multiplied by R_a, and the resulting signal is subtracted from the voltage reference that is provided from the current controller. The load W_L along with the local R_a feedback

© Springer International Publishing AG, part of Springer Nature 2018
S. N. Vukosavic, *Grid-Side Converters Control and Design*, Power Electronics
and Power Systems, https://doi.org/10.1007/978-3-319-73278-7_7

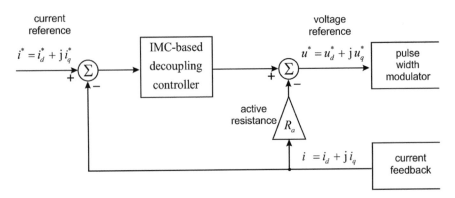

Fig. 7.1 Introduction of the active resistance feedback

can be replaced by the new, equivalent load W_{LRA}. The pulse transfer function $W_{LRA}(z)$ largely depends on the feedback acquisition chain, which can be either the one with the synchronous center-pulse sampling or the one with the one-PWM-period averaging. It also depends on the scheduling of the control tasks. In Sect. 7.1.1, the equivalent load W_{LRA} is derived for the conventional scheduling and the synchronous center-pulse sampling. In Sect. 7.1.2, the equivalent load W_{LRA} is derived for the conventional scheduling and the one-PWM-period averaging scheme. In Sect. 7.1.3, the equivalent load W_{LRA} is derived for the advanced scheduling with one-PWM-period averaging.

7.1.1   Equivalent Load with Synchronous Sampling

The pulse transfer function of the load $W_L(z)$ is given in Fig. 7.2a for the system with conventional schedule and synchronous sampling. The feedback signal $i^{FB}(z)$ is equal to the output current $i^e(z)$. The pulse transfer function $W_L(z)$ is obtained from (5.7) and given in (5.8). The load transfer function in the z-domain is obtained by dividing the complex images $i^e(z)$ and $u^e(z)$ in conditions where the disturbance $e^e(z)$ is equal to zero. From (5.8),

$$W_L(z) = \left. \frac{i^e(z)}{u^e(z)} \right|_{e^e=0} = \frac{T_S}{L} \frac{1}{z \cdot e^{j\omega_e T_S}\left(z \cdot e^{j\omega_e T_S} - e^{-\beta}\right)}. \tag{7.1}$$

In Fig. 7.2b, the equivalent pulse transfer function $W_{LRA}(z)$ replaces the block which comprises the original load and the active resistance feedback. The pulse transfer function W_{LRA} obtained by introducing (7.1) into (7.2) is given in (7.3). The result of (7.3) and the block diagram of Fig. 7.2b can be used as the starting point for the design of the decoupled digital current controller, based on the internal model principles.

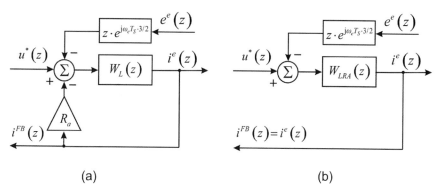

Fig. 7.2 Equivalent load with synchronous sampling and conventional schedule

$$W_{\text{LRA}}(z) = \frac{W_{\text{L}}(z)}{1 + W_{\text{L}}(z) \cdot R_{\text{a}}}. \tag{7.2}$$

$$W_{\text{LRA}}(z) = \frac{T_{\text{S}}}{L} \frac{1}{z^2 \cdot e^{2j\omega_e T_{\text{S}}} - z \cdot e^{j\omega_e T_{\text{S}}} \cdot e^{-\beta} + \frac{R_{\text{a}} T_{\text{S}}}{L}}. \tag{7.3}$$

7.1.2   Equivalent Load with One-PWM-Period Averaging

The pulse transfer function of the load $W_{\text{L}}(z)$ is given in Fig. 7.3a for the system with conventional schedule and with the oversampling-based one-PWM-period, explained in Sect. 3.3. The feedback signal $i^{\text{FB}}(z)$ is obtained by processing the output current $i^e(z)$ through the pulse transfer function $W_{\text{FB}}(z)$ of (3.32). The pulse transfer function $W_{\text{L}}(z)$ is given in (7.1). In Fig. 7.3b, the equivalent pulse transfer function $W_{\text{LRA}}(z)$ replaces the block which comprises the original load and the active resistance feedback.

$$W_{\text{LRA}}(z) = \frac{W_{\text{L}}(z)}{1 + W_{\text{L}}(z) \cdot W_{\text{FB}}(z) \cdot R_{\text{a}}}. \tag{7.4}$$

By introducing (7.1) and (3.32) into (7.4), one obtains the equivalent pulse transfer function of the load, given in (7.5). The result of (7.5) and the block diagram of Fig. 7.3b can be used as the starting point for the design of the decoupled digital current controller, based on the internal model principles.

$$W_{\text{LRA}}(z) = \frac{T_{\text{S}}}{L} \frac{z^2}{z^4 \cdot e^{2j\omega_e T_{\text{S}}} - z^3 \cdot e^{j\omega_e T_{\text{S}}} \cdot e^{-\beta} + z^2 \cdot \frac{R_{\text{a}} T_{\text{S}}}{4L} + z \cdot \frac{R_{\text{a}} T_{\text{S}}}{2L} + \frac{R_{\text{a}} T_{\text{S}}}{4L}}. \tag{7.5}$$

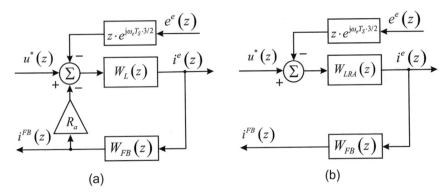

Fig. 7.3 Equivalent load with the one-PWM-period feedback averaging

7.1.3 Equivalent Load with the Advanced Scheduling

The pulse transfer function of the load $W_L(z)$ is given in Fig. 7.4a for the system with the advanced scheduling, developed in Chap. 6, and with the oversampling-based one-PWM-period, explained in Sect. 3.3. The feedback signal $i^{FB}(z)$ is obtained by processing the output current $i^e(z)$ through the pulse transfer function $W_{FB}(z)$ of (3.32). The pulse transfer function of the load $W_L(z)$, obtained with the advanced scheduling, is given in (7.6). In Fig. 7.4b, the equivalent pulse transfer function $W_{LRA}(z)$ replaces the block which comprises the original load and the active resistance feedback.

$$W_L(z) = \left.\frac{i^e(z)}{u^e(z)}\right|_{e^e=0} = \frac{T_S}{L}\frac{1}{z \cdot e^{j\omega_e T_S} - e^{-\beta}}. \tag{7.6}$$

The equivalent pulse transfer function of the load is

$$W_{LRA}(z) = \frac{W_L(z)}{1 + W_L(z) \cdot W_{FB}(z) \cdot R_a}. \tag{7.7}$$

By introducing (7.6) and (3.32) into (7.7), one obtains the equivalent pulse transfer function of the load, given in (7.8). The result of (7.8) and the block diagram of Fig. 7.4b can be used as the starting point for the design of the decoupled digital current controller, based on the internal model principles.

$$W_{LRA}(z) = \frac{T_S}{L} \cdot \frac{z^2}{z^3 e^{j\omega_e T_S} + z^2\left(\frac{R_a T_S}{4L} - \beta\right) + z\frac{R_a T_S}{2L} + \frac{R_a T_S}{4L}}. \tag{7.8}$$

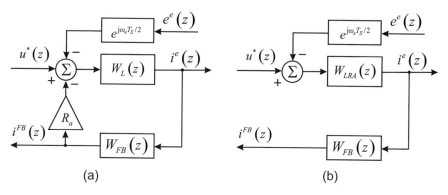

Fig. 7.4 Equivalent load with the advanced scheduling and feedback averaging

7.1.4 The Range of Stable R_a Gains

The analysis of Sect. 4.5 shows that the disturbance rejection capability of the digital current controller with active resistance feedback increases with the gain R_a. Thus, it is of interest to explore the range of applicable gains with the equivalent load transfer functions of (7.3), (7.5), and (7.8).

The pulse transfer function (7.3) represents the case with synchronous center-pulse sampling and with the conventional scheduling of the control tasks. The characteristic polynomial in denominator of (7.3) comprises the factor $R_a T_S/L$. The poles of the transfer function (7.3) are real until the gain reaches $R_a T_S/L = 0.247$. Stability limit of (7.3) is reached for $R_a T_S/L = 1$.

The pulse transfer function (7.5) represents the case with one-PWM-period feedback averaging and with the conventional scheduling of the control tasks. The characteristic polynomial in the denominator of (7.5) comprises the factor $R_a T_S/L$. There are four poles of the transfer function. At least one pair of poles is conjugate complex pair. The dominant poles are real until the gain reaches $R_a T_S/L = 0.138$. With $R_a T_S/L = 0.138$, the remaining pair of conjugate complex poles is at considerably larger frequency, and it has a damping of 0.5. Stability limit of (7.5) is reached for $R_a T_S/L = 0.68$.

The pulse transfer function (7.8) represents the case with one-PWM-period feedback averaging and with the advanced scheduling of the control tasks. Denominator of (7.8) comprises the factor $R_a T_S/L$, and there are three poles of the transfer function. All the poles are real until the gain reaches $R_a T_S/L = 0.223$. Stability limit of (7.8) is reached for $R_a T_S/L = 1.33$.

Thus, the largest range of applicable active resistance gains is reached with the case with one-PWM-period feedback averaging and with the advanced scheduling of the control tasks.

7.2 Design of Decoupling Controllers

With added active resistance feedback, the equivalent pulse transfer function of the load changes and assumes the form $W_{\mathrm{LRA}}(z)$, derived in the previous section. In order to provide the decoupled operation of the current controller, and to avoid the impact of the R_{a} gain on the input step response, it is necessary to apply the internal model control approach and to design the current controller which provides the decoupled step response unaffected by R_{a} gain. In Sect. 7.2.1, the decoupling current controller is designed for the equivalent pulse transfer function of the load W_{LRA} developed in Sect. 7.1.1, with the conventional scheduling and the synchronous center-pulse sampling. In Sect. 7.2.2, the decoupling current controller is designed for the equivalent pulse transfer function of the load W_{LRA} developed in Sect. 7.1.2, with the conventional scheduling and the one-PWM-period averaging scheme. In Sect. 7.2.3, the decoupling current controller is designed for the equivalent pulse transfer function of the load W_{LRA} developed in Sect. 7.1.3, with the advanced scheduling and with one-PWM-period averaging.

7.2.1 Conventional Scheduling with Synchronous Sampling

For the current controller with the conventional scheduling and synchronous center-pulse sampling, the equivalent pulse transfer function of the load is given in (7.3). An attempt to invert the dynamics of the load and to multiply the resulting pulse transfer function by the integrator $\alpha_{\mathrm{D}}z/(z - 1)$ results in (7.9). Attempted current controller $W_{\mathrm{REGX}}(z)$ has a numerator of the third order and a denominator of the first order. This implies prediction of two sampling intervals. Therefore, the practical current controller is obtained by dividing $W_{\mathrm{REGX}}(z)$ by z^2 (7.10).

$$W_{\mathrm{REGX}}(z) = \frac{1}{W_{\mathrm{LRA}}(z)} \cdot \frac{\alpha_{\mathrm{D}}z}{z-1} = \frac{\alpha_{\mathrm{D}}L \cdot z}{T_{\mathrm{S}}} \cdot \frac{z^2 \cdot e^{2j\omega_e T_{\mathrm{S}}} - z \cdot e^{j\omega_e T_{\mathrm{S}}} \cdot e^{-\beta} + \frac{R_{\mathrm{a}}T_{\mathrm{S}}}{L}}{z-1}.$$

$$(7.9)$$

$$W_{\mathrm{REG}}(z) = \frac{\alpha_{\mathrm{D}}z}{W_{\mathrm{LRA}}(z)(z-1)}$$

$$\cdot \frac{1}{z^2} = \frac{\alpha_{\mathrm{D}}L}{T_{\mathrm{S}}} \cdot \frac{z^2 \cdot e^{2j\omega_e T_{\mathrm{S}}} - z \cdot e^{j\omega_e T_{\mathrm{S}}} \cdot e^{-\beta} + \frac{R_{\mathrm{a}}T_{\mathrm{S}}}{L}}{z \cdot (z-1)}. \qquad (7.10)$$

The current controller of (7.10) is shown in Fig. 7.5. The closed-loop transfer function, calculated from the block diagram of Fig. 7.6, is given in (7.11), while the disturbance rejection pulse transfer function $Y^e(z)$ is given in (7.12).

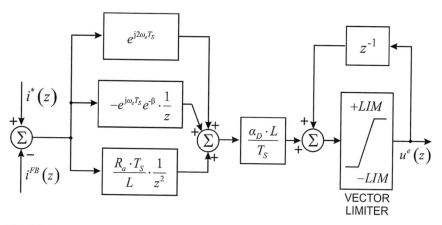

Fig. 7.5 Current controller of (7.10)

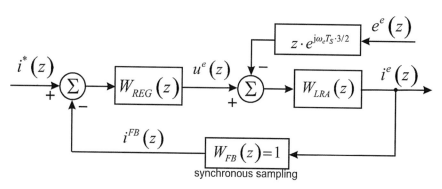

Fig. 7.6 Closed-loop current controller with conventional scheduling, with synchronous sampling, with active resistance feedback, and with the current controller W_{REG} of (7.10)

$$W_{\text{CL}}(z) = \left.\frac{i^e(z)}{i^*(z)}\right|_{e^e=0} = \frac{W_{\text{REG}}(z) \cdot W_{\text{LRA}}(z)}{1 + W_{\text{REG}}(z) \cdot W_{\text{LRA}}(z)} = \frac{\alpha_{\text{D}}}{z^2 - z + \alpha_{\text{D}}}. \qquad (7.11)$$

$$Y^e(z) = \frac{z \cdot e^{j\frac{3}{2}\omega_e T_S} W_{\text{LRA}}(z)}{1 + W_{\text{LRA}}(z) \cdot W_{\text{REG}}(z)}$$

$$= \frac{T_S}{L} \frac{z^2(z-1) \cdot e^{j\frac{3}{2}\omega_e T_S}}{(z^2 - z + \alpha_{\text{D}})\left(z^2 \cdot e^{2j\omega_e T_S} - z \cdot e^{j\omega_e T_S} \cdot e^{-\beta} + \frac{R_a T_S}{L}\right)}. \qquad (7.12)$$

7.2.2 Conventional Scheduling with Feedback Averaging

For the current controller with the conventional scheduling and with one-PWM-period averaging, the equivalent pulse transfer function of the load is given in (7.5). An attempt to invert the dynamics of the load and to multiply the resulting pulse transfer function by the integrator $\alpha_D z/(z-1)$ results in (7.13). Attempted current controller $W_{REGX}(z)$ has a numerator of the fourth order and a denominator of the second order. This implies prediction of two sampling intervals. Therefore, the practical current controller is obtained by dividing $W_{REGX}(z)$ by z^2 (7.14).

$$
\begin{aligned}
W_{REGX}(z) &= \frac{1}{W_{LRA}(z)} \cdot \frac{\alpha_D z}{z-1} \\
&= \frac{\alpha_D L}{T_S} \cdot \frac{z^4 \cdot e^{2j\omega_e T_S} - z^3 \cdot e^{j\omega_e T_S} \cdot e^{-\beta} + z^2 \cdot \frac{R_a T_S}{4L} + z \cdot \frac{R_a T_S}{2L} + \frac{R_a T_S}{4L}}{z \cdot (z-1)}.
\end{aligned}
$$

(7.13)

$$
\begin{aligned}
W_{REG}(z) &= \frac{1}{W_{LRA}(z)} \cdot \frac{\alpha_D z}{z-1} \cdot \frac{1}{z^2} = \frac{\alpha_D L}{T_S} \cdot \\
&\frac{z^4 \cdot e^{2j\omega_e T_S} - z^3 \cdot e^{j\omega_e T_S} \cdot e^{-\beta} + z^2 \cdot \frac{R_a T_S}{4L} + z \cdot \frac{R_a T_S}{2L} + \frac{R_a T_S}{4L}}{z^3 \cdot (z-1)}.
\end{aligned}
$$

(7.14)

The current controller of (7.14) is shown in Fig. 7.7. The closed-loop transfer function, calculated from the block diagram of Fig. 7.8, is given in (7.15), while the disturbance rejection pulse transfer function $Y^e(z)$ is given in (7.16).

$$
\begin{aligned}
W_{CL}(z) &= \left. \frac{i^e(z)}{i^*(z)} \right|_{e^e=0} = \frac{W_{REG}(z) \cdot W_{LRA}(z)}{1 + W_{REG}(z) \cdot W_{LRA}(z) \cdot W_{FB}(z)} \\
&= \frac{4\alpha_D z^2}{4z^4 - 4z^3 + \alpha_D z^2 + 2\alpha_D z + \alpha_D}.
\end{aligned}
$$

(7.15)

Disturbance pulse transfer function is

$$
Y^e(z) = z \cdot \frac{e^{j\frac{3}{2}\omega_e T_S} W_{LRA}(z)}{1 + W_{LRA}(z) \cdot W_{REG}(z) \cdot W_{FB}(z)} = \frac{T_S}{L} 4z^6(z-1) \cdot \frac{e^{j\frac{3}{2}\omega_e T_S}}{f_1(z) \cdot f_2(z)},
$$

(7.16)

where

$$
\begin{aligned}
f_1(z) &= 4z^4 - 4z^3 + \alpha_D z^2 + 2\alpha_D z + \alpha_D, \\
f_2(z) &= z^4 \cdot e^{2j\omega_e T_S} - z^3 \cdot e^{j\omega_e T_S} \cdot e^{-\beta} + z^2 \cdot \frac{R_a T_S}{4L} + z \cdot \frac{R_a T_S}{2L} + \frac{R_a T_S}{4L}.
\end{aligned}
$$

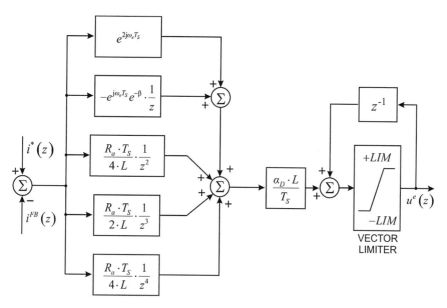

Fig. 7.7 Current controller of (7.14)

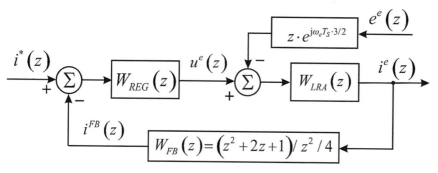

Fig. 7.8 Closed-loop current controller with conventional scheduling, with one-PWM-period feedback averaging, with active resistance feedback, and with the current controller W_{REG} of (7.14)

7.2.3 Advanced Scheduling with Feedback Averaging

For the current controller with the advanced scheduling and with one-PWM-period averaging, the equivalent pulse transfer function of the load is given in (7.8). An attempt to invert the dynamics of the load and to multiply the resulting pulse transfer function by the integrator $\alpha_D z/(z - 1)$ results in (7.17). Attempted current controller $W_{\text{REGX}}(z)$ has a numerator of the third order and a denominator of the second order. This implies prediction of one sampling interval. Therefore, the practical current controller is obtained by dividing $W_{\text{REGX}}(z)$ by z (7.18).

$$W_{\text{REGX}}(z) = \frac{1}{W_{\text{LRA}}(z)} \cdot \frac{\alpha_D z}{z - 1}$$

$$= \frac{\alpha_D L}{T_S} \cdot \frac{z^3 e^{j\omega_e T_S} + z^2 \left(\frac{R_a T_S}{4L} - \beta \right) + z \frac{R_a T_S}{2L} + \frac{R_a T_S}{4L}}{z \cdot (z - 1)}. \tag{7.17}$$

$$W_{\text{REG}}(z) = \frac{1}{W_{\text{LRA}}(z)} \cdot \frac{\alpha_D z}{z - 1} \cdot \frac{1}{z}$$

$$= \frac{\alpha_D L}{T_S} \cdot \frac{z^3 e^{j\omega_e T_S} + z^2 \left(\frac{R_a T_S}{4L} - \beta \right) + z \frac{R_a T_S}{2L} + \frac{R_a T_S}{4L}}{z^2 \cdot (z - 1)}. \tag{7.18}$$

The current controller of (7.18) is shown in Fig. 7.9. The closed-loop transfer function, calculated from the block diagram of Fig. 7.10, is given in (7.19), while the disturbance rejection pulse transfer function $Y^e(z)$ is given in (7.20).

$$W_{\text{CL}}(z) = \left. \frac{i^e(z)}{i^*(z)} \right|_{e^e = 0} = \frac{4\alpha_D z^2}{4z^3 + (\alpha_D - 4)z^2 + 2\alpha_D z + \alpha_D}. \tag{7.19}$$

Disturbance pulse transfer function is

$$Y^e(z) = \frac{e^{j \frac{\omega_e T_S}{2}} W_{\text{LRA}}(z)}{1 + W_{\text{LRA}}(z) W_{\text{REG}}(z) W_{\text{FB}}(z)} = \frac{T_S}{L} \frac{4z^4 (z - 1) \cdot e^{j \frac{\omega_e T_S}{2}}}{f_1(z) \cdot f_2(z)}, \tag{7.20}$$

where

$$f_1(z) = 4z^3 + (\alpha_D - 4)z^2 + 2\alpha_D z + \alpha_D,$$
$$f_2(z) = z^3 e^{j\omega_e T_S} + z^2 \left(\frac{R_a T_S}{4L} - \beta \right) + z \frac{R_a T_S}{2L} + \frac{R_a T_S}{4L}.$$

7.3 Disturbance Suppression in Synchronous Frame

In Sect. 7.2, the closed-loop transfer functions $W_{\text{CL}}(z)$ and the disturbance rejection transfer functions were derived for the equivalent load W_{LRA} is derived with the conventional scheduling and the synchronous center-pulse sampling, for the equivalent load W_{LRA} is derived with the conventional scheduling and the one-PWM-period averaging scheme, and for the equivalent load W_{LRA} is derived with the advanced scheduling with one-PWM-period averaging. The relevant closed-loop transfer functions are the same as the ones obtained in Chaps. 5 and 6, for the corresponding current controllers that do not use the active resistance feedback. Thus, both the gain setting procedures and the analysis of the closed-loop performances as regarding the input step response are already dealt with before. In this section, the attention is focused on the disturbance rejection capability. Disturbance

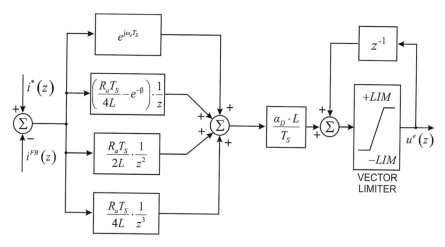

Fig. 7.9   Current controller of (7.18)

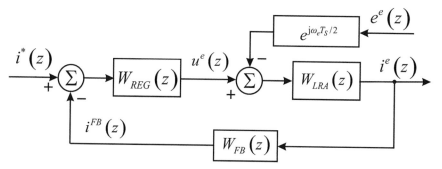

Fig. 7.10   Closed-loop current controller with advanced scheduling, with one-PWM-period feedback averaging, with active resistance feedback, and with the current controller W_{REG} of (7.18)

transfer function developed in (7.12), (7.16), and (7.20) are derived in d-q synchronous coordinate frame. They provide the response of the output current $i^e(z)$ to the changes of the voltage disturbance $e^e(z)$ in d-q frame. Disturbance rejection is studied by exploring the impact of the active resistance gain R_a on the response in time domain, as well as on the integral of the output current error. The outcome of this study is the suggested parameter setting for the active resistance gain.

7.3.1   The Applicable Range of R_a Gains

The range of stable R_a gains is studied in Sect. 7.1.4. Larger values of R_a contribute to an improved suppression of the voltage disturbances. In Table 7.1, the values of the vector margin are given for the equivalent load pulse transfer functions (7.3),

Table 7.1 The vector margin for the equivalent load pulse transfer functions of (7.3), (7.5), and (7.8)

Gain $R_a T_S/L$	W_{LRA} of (7.3)	W_{LRA} of (7.5)	W_{LRA} of (7.8)
0.1	0.8746	0.8086	0.8832
0.2	0.7619	0.6455	0.7838
0.3	0.6557	0.4968	0.6936
0.4	0.5541	0.3581	0.6100
0.5	0.4559	0.2273	0.5316
0.6	0.3606	0.1027	0.4574

(7.5), and (7.8), derived in Sects. 7.1.1, 7.1.2, and 7.1.3. When the vector margin drops below 0.5, the dynamic response of the system is poorly damped and more sensitive to parameter changes. In the case with conventional scheduling and with the synchronous sampling, the vector margin remains above 0.5 of the gains $R_a T_S/L$ that do not exceed 0.4. In the case with conventional scheduling and with one-PWM-period feedback averaging, synchronous sampling, the vector margin remains above 0.5 of the gains $R_a T_S/L$ that do not exceed 0.2. In the case with advanced scheduling and one-PWM-period feedback averaging, the vector margin remains above 0.5 of the gains $R_a T_S/L$ that do not exceed 0.5.

7.3.2   Simulation of the Dynamic Response

Dynamic response to the step voltage disturbance is simulated in Matlab. The Matlab script used to generate the responses is given in Table 7.2. Simulation considers both the impact of the d-axis disturbance to the d-axis current, as well as the impact of the d-axis disturbance to the q-axis current.

In Table 7.2, the Matlab script derives the response of the current controller with conventional scheduling and synchronous sampling. The remaining two cases are simulated in a like manner, by introducing the appropriate expressions for the numerator and denominator of $Y^e(z)$.

7.3.3   Dynamic Response

The output response to the step change of the voltage disturbance is obtained by simulating the system with $R = 0.47$ Ω, $L = 3.8$ mH, $T_S = 50$ μs, for the output frequency of $f_e = 50$ Hz, and for the gain of $\alpha_D = 0.3$. The active resistance gain ($R_a T_S/L$) is changed from 0.1 up to 0.5.

In Fig. 7.11, the traces represent the disturbance rejection response obtained with the equivalent pulse transfer function of (7.3), with conventional scheduling and synchronous sampling. The gain $R_a T_S/L$ is changed from 0.1 up to 0.5. The traces

Table 7.2 Matlab code which calculates dynamic response of the current controller to the step change of the voltage disturbance in *d-q* frame

```
close all; Rs = 0.47; L = 0.0038; T = 0.00005;    % Initialization of the system
b = exp(-Rs*T/L); we = 314; a = 0.3; ratl = 0.25; % parameters  and gains
       % conventional scheduling syncronous sampling
figure; hold on; grid;
for j = 1:5,                                   % The gain changing from
ratl = j/10;                                   % RaTs/L = 0.1 up to 0.5
num = T/L*exp(1i*1.5*we*T)*[1 -1 0 0];         % Numerator of Ye
den1 = [1 -1 a];
den2 = [exp(1i*2*we*T)-exp(1i*we*T)*b ratl];
den = conv(den1, den2);                        % Denominator of Ye
rr = dstep(num,den,50);
stairs(real(rr)+j/25);                         % d-response to
axis([0 50 0.03 0.23])                         % d-disturbance
end
 figure; hold on; grid;
for j = 1:5,                                   % The gain changing from
ratl = j/10;                                   % RaTs/L = 0.1 up to 0.5
num = T/L*exp(1i*1.5*we*T)*[1-1 0 0];          % Numerator of Ye
den1 = [1-1 a];
den2 = [exp(1i*2*we*T) -exp(1i*we*T)*b ratl];
den = conv(den1, den2);                        % Denominator of Ye
rr = dstep(num,den,50);
stairs(imag(rr)+j/25);                         % q-response to
end                                            % d-disturbance
```

Fig. 7.11 Disturbance rejection response obtained with the equivalent pulse transfer function of (7.3), with conventional scheduling and synchronous sampling. The gain α_D is set to 0.3. The gain R_aT_S/L is changed from 0.1 up to 0.5. The traces represent the response of the *d*-axis current to the voltage step disturbance of the unit amplitude (1 V) in the *d*-axis

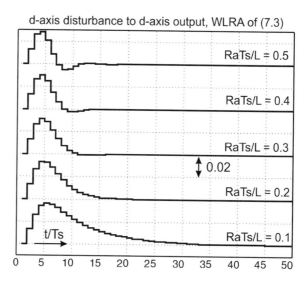

represent the response of the *d*-axis current to the voltage step disturbance of the unit amplitude (1 V) in the *d*-axis. The gain can reach 0.5 without oscillations. The peak of the output current error reaches 0.03, while duration of the error pulse is from six to seven sampling periods. For the same system, the same gain setting, and the same *d*-axis voltage disturbance, the traces of the *q*-axis current response are given in

Fig. 7.12 Disturbance rejection response obtained with the equivalent pulse transfer function of (7.3), with conventional scheduling and synchronous sampling. The gain α_D is set to 0.3. The gain R_aT_S/L is changed from 0.1 up to 0.5. The traces represent the response of the q-axis current to the voltage step disturbance of the unit amplitude (1 V) in the d-axis

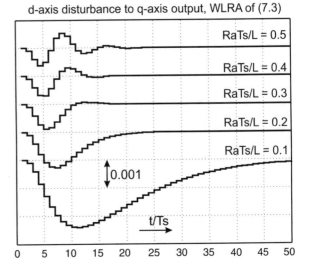

Fig. 7.12. The peak of the output current error reaches 0.0025 with the gain of 0.1, while the gain 0.5 contributes to significant oscillations.

The traces in Fig. 7.13 are obtained with the equivalent pulse transfer function of (7.5), with conventional scheduling and with one-PWM-period feedback averaging. The traces represent the response of the d-axis current to the voltage step disturbance of the unit amplitude (1 V) in the d-axis. The peak value of the current error is 0.05, while the duration of the current error pulse is from 12 to 16 sampling periods. For the gains larger than 0.2, the waveforms include significant oscillations. For the same system, the same gain setting, and the same d-axis voltage disturbance, the traces of the q-axis current response are given in Fig. 7.14. The peak of the output current error reaches 0.003, while the gains larger than 0.2 introduce oscillations.

The traces in Figs. 7.15 and 7.16 are obtained with the equivalent pulse transfer function of (7.8), with advanced scheduling and with one-PWM-period feedback averaging. The gain R_aT_S/L can reach 0.5 without oscillations. The d-axis current error remains lower than 0.025, with a duration of up to six sampling intervals. In Fig. 7.16, the q-axis waveforms have the peak values lower than 0.001. Considering time-domain responses, the best disturbance rejection is obtained with the advanced scheduling scheme and with one-PWM-period averaging.

7.4   Disturbance Suppression in Stationary Frame

Disturbance transfer functions (7.12), (7.16), and (7.20) define the response of the load current i^e in d-q frame to the voltage disturbance $-e^e$ in the same d-q frame. In practical grid-side inverters, the voltage disturbances originate from the grid, and they reside in the stationary α-β coordinate frame, as well as the response of the

Fig. 7.13 Disturbance rejection response obtained with the equivalent pulse transfer function of (7.5), with conventional scheduling and with one-PWM-period feedback averaging. The gain α_D is set to 0.3. The gain $R_a T_S/L$ is changed from 0.1 up to 0.5. The traces represent the response of the d-axis current to the voltage step disturbance of the unit amplitude (1 V) in the d-axis

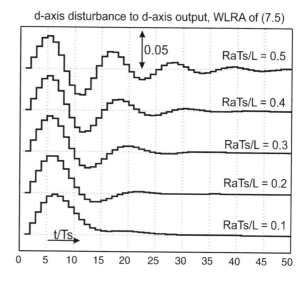

Fig. 7.14 Disturbance rejection response obtained with the equivalent pulse transfer function of (7.5), with conventional scheduling and with one-PWM-period feedback averaging. The gain α_D is set to 0.3. The gain $R_a T_S/L$ is changed from 0.1 up to 0.5. The traces represent the response of the q-axis current to the voltage step disturbance of the unit amplitude (1 V) in the d-axis

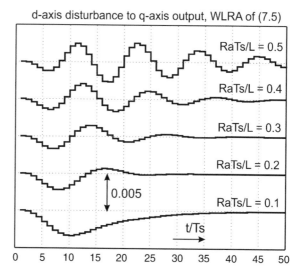

output currents to such disturbances. For this reason, it is of interest to obtain the transfer function $Y^s = i^s/(-e^s)$ that defines the output current response i^s in the stationary frame to voltage disturbances in the stationary frame. The transfer functions defined in d-q synchronous frame can be used to obtain the corresponding transfer functions in the stationary frame, using the approach outlined in expressions (4.29) and (4.30). Considering the stationary frame voltage disturbance at the frequency ω_x, and assuming that the synchronous d-q frame revolves at the speed ω_e, Y^s can be obtained from Y^e,

Fig. 7.15 Disturbance
rejection response obtained
with the equivalent pulse
transfer function of (7.8),
with advanced scheduling
and with one-PWM-period
feedback averaging. The
gain α_D is set to 0.3. The
gain R_aT_S/L is changed from
0.1 up to 0.5. The traces
represent the response of the
d-axis current to the voltage
step disturbance of the unit
amplitude (1 V) in the d-axis

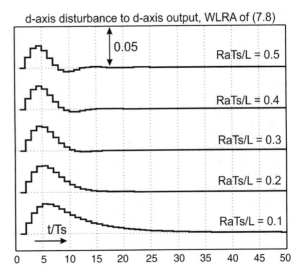

Fig. 7.16 Disturbance
rejection response obtained
with the equivalent pulse
transfer function of (7.8),
with advanced scheduling
and with one-PWM-period
feedback averaging. The
gain α_D is set to 0.3. The
gain R_aT_S/L is changed from
0.1 up to 0.5. The traces
represent the response of the
q-axis current to the voltage
step disturbance of the unit
amplitude (1 V) in the d-axis

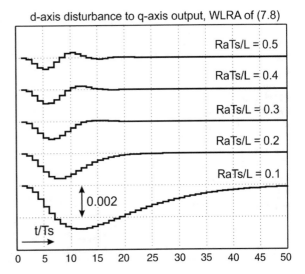

$$Y^s(z) = Y^s\left(e^{j\omega_x T_S}\right) = Y^e\left(e^{j\omega_x T_S}e^{-j\omega_e T_S}\right) = Y^e\left(e^{j(\omega_x - \omega_e)T_S}\right). \qquad (7.21)$$

Thus, when considering the stationary frame voltage disturbance at the frequency
ω_x and with the amplitude U_x, the amplitude I_x of the load current response is
obtained as Y^sU_x, where Y^s is obtained from Y^e by replacing the frequency ω_x by $\omega_x -$
ω_e. The Matlab script that calculates the frequency characteristics of the pulse
transfer functions derived from (7.12), (7.16), and (7.20).

7.4.1   The Frequency Characteristic of $\mathbf{Y}^s$ with $\mathbf{R}_a = 0$

The frequency response of the pulse transfer functions Y^s, developed from (7.12), (7.16), and (7.20) in the case when the active resistance gain is set to zero, is given in Fig. 7.17, with the gain α_D which is set to 0.3 and with the remaining parameters of the system defined in the first several lines of the script given in Table 7.3. The same plot is given again in Fig. 7.18, with the horizontal axis adjusted so as to show the disturbance rejection of the most important low order line harmonics.

In Fig. 7.18, the frequency characteristic is not symmetrical. Namely, the left part of the plot, corresponding to *negative frequencies*, has larger values of Y^s than the right side of the plot, the one that corresponds to *positive frequencies*. Although it is quite unusual to take into consideration the negative frequencies, it makes sense in the context of the frequency characteristic in Fig. 7.18. In three-phase systems, the low order line harmonics such as the fifth could create the vectors that revolve in positive direction (direct-sequence fifth harmonic) or the vectors that revolve in negative direction (inverse-sequence fifth harmonic). From Fig. 7.18, disturbance rejection is better for direct-sequence line harmonics. For the direct-sequence fifth and seventh harmonics, Y^s remains below 0.04. On the other hand, the inverse-sequence fifth harmonic encounters $Y^s \approx 0.055$. From Fig. 7.18, one concludes that the frequency characteristic of Y^s remains roughly the same for all the three current controllers that were considered within the script of Table 7.3. The values of Y^s can be reduced considerably by the use of $R_a > 0$, which is analyzed in the subsequent section.

7.4.2   The Frequency Characteristic of $\mathbf{Y}^s$ with $\mathbf{R}_a > 0$

The frequency response of the pulse transfer functions Y^s, developed from (7.12), (7.16), and (7.20), is given in Figs. 7.19, 7.20, and 7.21. The gain α_D is set to 0.3, while the gain R_aT_S/L is changed within the range that guarantees the vector margin larger than 0.5.

In Fig. 7.19, the values of Y^s reach 0.05; in Fig. 7.20 they reach 0.18, while in Fig. 7.21 the values of Y^s reach 0.035. The best results are obtained with the system with advanced scheduling with one-PWM-period feedback averaging, with the pulse transfer function of (7.20) and with the frequency characteristic of Y^s given in Fig. 7.21. With the gain $R_aT_S/L = 0.5$, the vector margin remains above 0.5, while the maximum value of Y^s remains below 0.025.

The proper interpretation of the results $|Y^s(j\omega)| < 0.025$ needs an additional discussion. Notice in Table 7.3 that the results in this section were obtained with the system where the load inductance has the value of $L = 3.8$ mH. At the line frequency of $f_e = 50$ Hz, the corresponding reactance is $\omega_eL = 1.2\ \Omega$. From the discussion on the design of the *LCL* filters in Chap. 2, it is reasonable to assume that the equivalent series reactance of the filter reaches 10% of the rated impedance.

Fig. 7.17 Frequency
dependence of the
disturbance rejection
transfer function Y^s in the
stationary frame, obtained
with $R_a = 0$ from the pulse
transfer functions Y^e is given
in (7.12), (7.16), and (7.20).
The gain is $\alpha_D = 0.3$

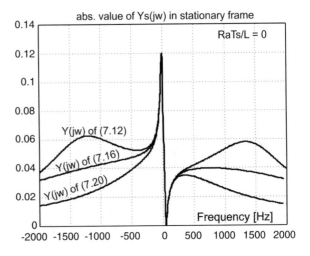

Table 7.3 Matlab code which calculates the frequency characteristic of the disturbance transfer function in the stationary coordinate frame

```
close all; clear all; Rs = 0.47; L = 0.0038; T = 0.00005; b = exp(-Rs*T/L); a = 0.3;
we = 314;   % "we" is the speed of the d-q frame in [rad/s], 314 --> 50 Hz
for i = 1:4000,
   fff = i-2000; www = 2*pi*fff; ff(i)=fff; % Calculate the frequency vector and zz,
   zz(i) = exp(1i*T*(www-we));              %  -2000 Hz  <  fff  <  +2000 Hz
end;
       % conventional scheduling syncronous sampling
figure; hold on; grid;
for jj = 1:4,                               % sweep the gains from 0.1 up to 0.4
ratl = jj/4*0.4;
num = T/L*(zz.^3 -zz.^2)*exp(1i*1.5*we*T); % Calculate the pulse
den1 = (zz.*zz -zz +a);                     %   t.f. according to (7.12)
den2 = zz.*zz*exp(1i*2*we*T) -exp(1i*we*T)*b*zz + ratl;
Y = num./den1./den2;  plot(ff,abs(Y));
end
     % conventional scheduling with feedback averaging, calculate the transfer
     % function according to the expression (7.16)
 figure; hold on; grid;
for jj = 1:4,
ratl = jj/4*0.2;
num = 4*T/L*exp(1i*1.5*we*T)*(zz.^6).*(zz-1);
den1 = 4*zz.^4 -4*zz.^3 +a*zz.^2 + 2*a*zz + a;
den2 = (zz.^4)*exp(1i*2*we*T) -exp(1i*we*T)*b*zz.^3+ratl/4*zz.^2+ratl/2*zz+ratl;
Y = num./den1./den2;  plot(ff,abs(Y));
end
     % Advanced scheduling with feedback averaging, calculate the transfer
     % function according to the expression (7.20)
figure; hold on; grid;
for jj = 1:4,
ratl = jj/4*0.5;
num = 4*T/L*exp(1i*0.5*we*T)*(zz.^4).*(zz-1);
den1 = 4*zz.^4 + (a-4)*zz.^2 +2*a*zz + a;
den2 = (zz.^3)*exp(1i*we*T)+(ratl/4-b)*zz.^2 + ratl/2*zz +ratl/4;
Y = num./den1./den2;  plot(ff,abs(Y));
End
```

Fig. 7.18 The frequency dependence of the previous figure is plotted again, with the frequency axis adjusted to cover the most relevant low order line harmonics

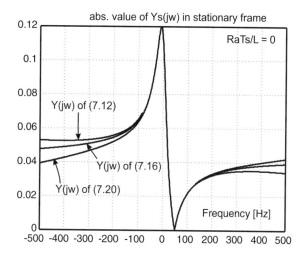

Fig. 7.19 Frequency dependence of the disturbance rejection transfer function Y^s in the stationary frame. The corresponding pulse transfer function Y^e is given in (7.12), and it is obtained with the conventional scheduling and synchronous sampling. The gain α_D is set to 0.3. The gain $R_a T_S/L$ is changed within the range [0.1...0.4], where the vector margin is larger than 0.5

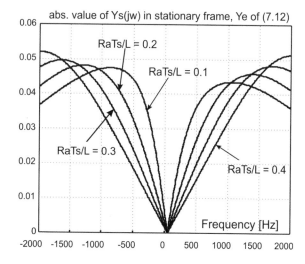

Thus, the rated impedance of the system considered within the script of Table 7.3 is $Z_n = 12 \ \Omega$, while the rated admittance is $Y_n = 1/Z_n = 0.0833$ S (Siemens).

The maximum value of $Y^s = 0.025$, reached at $f = 2000$ Hz in Fig. 7.21, corresponds to, roughly, 30% of the rated admittance. In practice, the voltage disturbance at 2 kHz, having an amplitude equal to 5% of the rated voltage, would provoke the current error at 2 kHz with an amplitude of $0.3 \cdot 5\% = 1.5\%$. Other than the disturbance rejection at 2 kHz, it is of interest to study the effectiveness of the active resistance gain in suppressing the low order line harmonics. The point that corresponds to the direct-sequence (that is, *positive direction*) fifth harmonic is denoted by (A) in Fig. 7.21. Compared to the results in Figs. 7.17 and 7.18, the admittance Y^s for the fifth harmonic is considerably reduced. At the operating point (A) in Fig. 7.21, the value of Y^s is close to 0.006, which is roughly 7.2% of the rated

Fig. 7.20 Frequency dependence of the disturbance rejection transfer function Y^s in the stationary frame. The corresponding pulse transfer function Y^e is given in (7.16), and it is obtained with the conventional scheduling and one-PWM-period averaging. The gain α_D is set to 0.3. The gain R_aT_S/L is changed within the range [0.1...0.2], where the vector margin is larger than 0.5

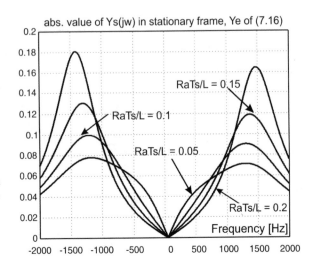

Fig. 7.21 Frequency dependence of the disturbance rejection transfer function Y^s in the stationary frame. The corresponding pulse transfer function Y^e is given in (7.20), and it is obtained with the advanced scheduling and one-PWM-period averaging. The gain α_D is set to 0.3. The gain R_aT_S/L is changed within the range [0.1...0.5], where the vector margin is larger than 0.5

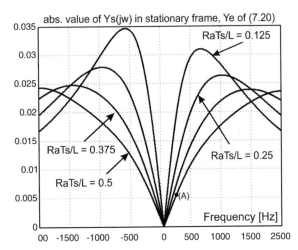

admittance. Thus, in cases where the line voltages have a direct-sequence fifth harmonic equal to 5% of the rated voltage, the consequential output current has the error at fifth harmonic which is lower than 0.36% of the rated current, the value that can easily be tolerated in most of the cases.

7.5 Experimental Results

In this Chapter, the digital current controller uses the decoupling action with relatively large gains αD, in order to achieve a high bandwidth input step response. At the same time, the controller uses the active resistance feedback with significant values of the local $R_a T_S/L$ gain, suited to suppress the impact of the voltage disturbances on the output current. The imperfections of the systems include the nonlinearity and the lockout time of the PWM voltage actuator, a finite resolution of the feedback acquisition system, a finite computation delays, and the usual system limitations such as the current limit, voltage limit, and the electromagnetic noise introduced by the PWM power conversion. The interaction of large gains with the system could be the source of undesired phenomena and performance degradation. Therefore, it is necessary to confirm the previous findings, obtained by the analysis and the computer simulation. This confirmation is performed experimentally, on the setup which is described in Sect. 5.4.1, comprising the PWM inverters, the DSP controller, and a brushless permanent magnet motor which is used as the load.

7.5.1 Parameters of the Experimental Setup

The key parameters of the experimental setup that is used in experimental verification are given in Sect. 5.4.1. The experiments were performed by introducing the step changes into the reference currents in d-q frame, and also by introducing forced step changes into the voltage references, in order to emulate the voltage disturbance. The waveforms contained in the subsequent plots represent the output currents in d-q frame, stored into RAM memory of the DSP controller. Thus, all the signals are obtained at the output of the feedback chain, and they correspond to $i^{FB}(z) = W_{FB}(z) \cdot i^e(z)$. The sampling time T_S is 50 μs, and the switching frequency is $f_{PWM} = 10$ kHz. The load parameters are $R = 0.47\ \Omega$ and $L = 3.4$ mH and the gain $\alpha_D = 0.3$. The current controller given in (7.18) and shown in Fig. 7.9 uses the advance scheduling and the one-PWM-period feedback averaging. The active resistance gain $(R_a T_S/L)$ is changed from 0.22, the value that results in real poles of the equivalent load, up to 0.54, the largest value that still keeps the vector margin above 0.5 (Table 7.1), while some tests were performed even with $R_a T_S/L$ approaching the stability limit of 1.33.

7.5.2 Input Step Response

The traces given in Fig. 7.22 correspond to the feedback signals i^{FB} of the d-axis and q-axis currents obtained during the input step response. The reference current for the d-axis is maintained constant, while the reference current for the q-axis exhibits a step change by 5 A. The experiment is performed with the current controller of (7.18), shown in Fig. 7.9, applying the advanced scheduling and the one-PWM-period feedback averaging. The output frequency is maintained at $f_e = 150$ Hz. There are three pairs of i_d-i_q current traces. One pair of traces is obtained with $R_a T_S/L = 0$. The other pair of traces is obtained with $R_a T_S/L = 0.22$, the largest value which results in real poles of the equivalent load W_{LRA}. The third pair of traces is obtained with $R_a T_S/L = 0.54$, the largest gain that maintains the vector margin

Fig. 7.22 The input step response with the current controller of (7.18), shown in Fig. 7.9, applying the advanced scheduling and the one-PWM-period feedback averaging. The output frequency is $f_e = 150$ Hz, the q-axis current exhibits a step of 5 A, while the gain $R_a T_S/L$ changes from 0 to 0.22 (real poles of W_{LRA}) and 0.54 (the vector margin VM > 0.5)

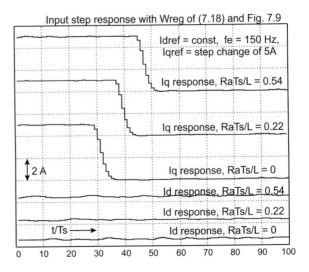

VM > 0.5. The experimental traces demonstrate that the step response remains unchanged by the introduction of the active resistance gain. At the same time, transient phenomena in the d-axis and q-axis remains decoupled. The character of the input step response, the rise time, and the settling time do not seem to change either.

The traces in Fig. 7.23 represent the step response of the q-axis current in cases where the active resistance gain is increased up to the levels that are close to the stability limit. The current controller is given in (7.18) and shown in Fig. 7.9. It applies the advanced scheduling and the one-PWM-period feedback averaging. The output frequency is $f_e = 270$ Hz, the q-axis current exhibits a step of 5 A, while the gain $R_a T_S/L$ changes from 0 to 0.54 (the vector margin VM > 0.5) and 1.21 (coming close to the stability limit of W_{LRA}, 1.33). The step response obtained with $R_a T_S/L = 0.54$ and with the vector margin VM > 0.5 (Table 7.1) remains the same as the step response obtained without the active resistance feedback. As the gain $R_a T_S/L$ approaches the stability limit of 1.33, the step response acquires small, poorly damped parasitic oscillations. Thus, the practicable values of the active resistance gain $R_a T_S/L$ are those that keep VM > 0.5.

7.5.3   Disturbance Rejection

The proper disturbance rejection test requires an introduction of the step change into the line voltage or, in the case where the load is an electrical machine, the step response into the back-electromotive force. Since the experimental setup is running with a brushless dc motor as the load, there is no viable way to introduce the step response into the actual back-electromotive force. Instead, the test was performed by introducing an appropriate replacement for such a step. The voltage reference

Fig. 7.23 The input step response with the current controller of (7.18), shown in Fig. 7.9, applying the advanced scheduling and the one-PWM-period feedback averaging. The output frequency is $f_e = 270$ Hz, the q-axis current exhibits a step of 5 A, while the gain $R_a T_S/L$ changes from 0 to 0.54 (the vector margin VM > 0.5) and 1.21 (coming close to the stability limit of W_{LRA}, 1.33)

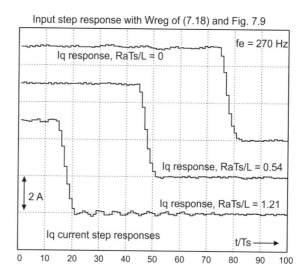

Input step response with Wreg of (7.18) and Fig. 7.9

provided by the current controller comes from the integrator, shown in the right-hand side in Fig. 7.9. By changing the value stored within the integrator in a stepwise manner, this advertent and forced step produces the effects equivalent to those encountered the hypothetical step change of the back-electromotive force.

In Fig. 7.24, the traces correspond to the feedback signals that represent the measured d-axis current and q-axis current. The test is performed with the current controller of (7.18), shown in Fig. 7.9, applying the advanced scheduling and the one-PWM-period feedback averaging. The output frequency is set to $f_e = 50$ Hz. The q-axis voltage, obtained at the output of the current controller, is changed in a stepwise manner by 67 V. This change is effectuated by software, by forcing the q-axis integrator within the current controller to change by the desired amount. The gain $R_a T_S/L$ is equal to zero. The transient current error does not decay over the time span of hundreds of sampling intervals, and it reaches the peak value of 3.5 A. Along with the q-axis transient, there is also a significant current error in d-axis.

The same test is repeated in Fig. 7.25 with a different value of the active resistance gain. The gain $R_a T_S/L$ is set to 0.22, the value which results in real poles of the equivalent load pulse transfer function W_{LRA}. The transient current error reaches the peak of roughly 2.2 A, while the transient ends in 15–20 sampling periods. Contemporary d-axis current error is an order of magnitude smaller than in the previous case.

Another test of the same kind is repeated in Fig. 7.26, with the gain $R_a T_S/L$ set to 0.54, the value which results in the vector margin VM > 0.5. The peak of the current error is reduced from 2.2 A down to 1.8 A, while the transient process ends in 12–15 sampling periods. Finally, the active resistance gain $R_a T_S/L$ is increased up to 0.81 in Fig. 7.27, which produces a further reduction of the peak current error down to 1.5 A. Yet, the experimental traces in Fig. 7.27 include significant poorly damped oscillations. Thus, based on the experimental evidence, one concludes that the practicable values of the active resistance gain remain limited to those that keep the vector margin above VM > 0.5 (Table 7.1).

Fig. 7.24 Disturbance step
response with the current
controller of (7.18), shown
in Fig. 7.9, applying the
advanced scheduling and
the one-PWM-period
feedback averaging. The
output frequency is
$f_e = 50$ Hz. The q-axis
voltage, obtained at the
output of the current
controller, changed by 67 V
by forcing the q-axis
integrator comprised within
the current controller. The
gain R_aT_S/L is equal to zero

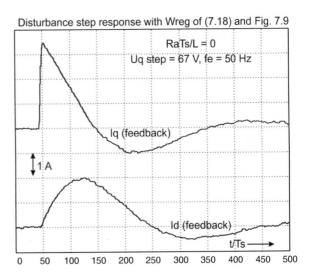

Fig. 7.25 Disturbance step
response with the current
controller of (7.18), shown
in Fig. 7.9, applying the
advanced scheduling and
the one-PWM-period
feedback averaging. The
output frequency is
$f_e = 50$ Hz. The q-axis
voltage, obtained at the
output of the current
controller, is changed by
67 V by forcing the q-axis
integrator comprised within
the current controller. The
gain R_aT_S/L is equal to 0.22,
the value which results in
real poles of the equivalent
load pulse transfer function
W_{LRA}

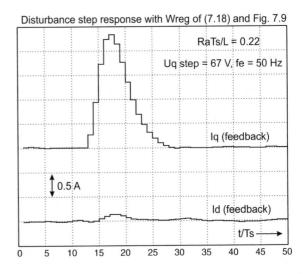

7.6   Concluding Remarks

Analytical considerations and the experimental evidence presented in this chapter
demonstrate that the application of active resistance feedback in conjunction with the
error-free sampling provides an efficient means of suppressing the voltage disturbances. The one-PWM-period feedback averaging is of particular importance, since
relatively large active resistance gains amplify not only the feedback but also the
noise and the sampling errors. However, the feedback averaging introduces an
additional transport delay. In order to cope with undesired couplings and transport

Fig. 7.26 Disturbance step response with the current controller of (7.18), shown in Fig. 7.9, applying the advanced scheduling and the one-PWM-period feedback averaging. The output frequency is $f_e = 50$ Hz. The q-axis voltage, obtained at the output of the current controller, is changed by 67 V by forcing the q-axis integrator comprised within the current controller. The gain $R_a T_S/L$ is equal to 0.54, the value which results in the vector margin VM > 0.5

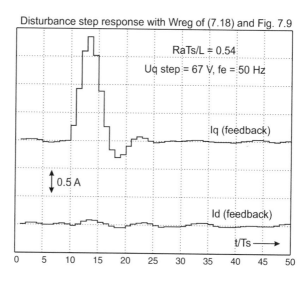

Fig. 7.27 Disturbance step response with the current controller of (7.18), shown in Fig. 7.9, applying the advanced scheduling and the one-PWM-period feedback averaging. The output frequency is $f_e = 50$ Hz. The q-axis voltage, obtained at the output of the current controller, is changed by 67 V by forcing the q-axis integrator comprised within the current controller. The gain $R_a T_S/L$ is equal to 0.81

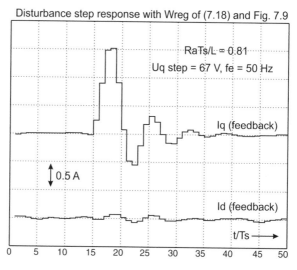

delays, it is necessary to apply the concept of the internal model control and to design the structure of the current controller which takes into account both the presence of the active resistance feedback within the equivalent load and the transport delays caused by the error-free feedback acquisition. The range of applicable active resistance gains is studied in detail, and this analysis provides the practical recommendation. The relative value of the active resistance gain can be set anywhere between $R_a T_S/L = 0.22$, the value which still provides real poles of the equivalent load $W_{LRA}(z)$, and $R_a T_S/L = 0.54$, the largest value that still provides the vector margin VM > 0.5. The applicable range of the active resistance feedback values is confirmed by simulations and experiments.

The experimental traces confirm that the input step response of the controller does not get affected by the active resistance feedback for all active resistance values up to the one that would bring the equivalent load (that is, the load with the active resistance feedback) close to the stability limit. The experiments confirm that the active resistance feedback can be used in conjunction with error-free one-PWM-period averaging while keeping the input step response decoupled while increasing the disturbance suppression capability by an order of magnitude.

In Chap. 5, the effects of the step voltage disturbance are quantified by the integral of the consequential output current error. In (5.21), this integral is expressed as $I = E_d/\alpha_D/R$, where E_d is the amplitude of the voltage step, α_D is the gain of the current controller, and R is the equivalent load resistance. In cases with the active resistance feedback, the expression should read $I = E_d/\alpha_D/(R + R_a)$, where R is the actual resistance of the load and R_a is the active resistance feedback. For the system considered in Matlab script of Table 7.2 and used in experimental verification, $R = 0.47\ \Omega$, $R_aT_S/L = 0.54$, $L = 0.038$, and $T_S = 50\ \mu s$. Thus, $R_a = 41\ \Omega$. In other words, the application of the active resistance feedback reduces the integral of the output current error roughly $R_a/R = 87$ times.

Question (7.1) Digital current control system with double update, center-pulse sampling and conventional task scheduling has the load transfer function given in expression (5.8). The sampling time is $T_S = 50\ \mu s$, the load inductance is $L = 10\ mH$, the angular frequency ω_e is close to zero, while the factor $\exp(-\beta)$ is close to 1. It is necessary to calculate the maximum stable value of the active resistance feedback R_a.

Answer (7.1) The load transfer function of the given system is introduced in Chap. 5.

$$W_L(z) = \left.\frac{i^e(z)}{u^e(z)}\right|_{e^e=0} = \frac{T_S}{L}\frac{1}{z \cdot e^{j\omega_eT_S}(z \cdot e^{j\omega_eT_S} - e^{-\beta})}$$

With double update rate, center sampling, and the active resistance feedback, the modified load pulse transfer function is

$$W_{LRA}(z) = \frac{\dfrac{T_S}{L}}{z \cdot e^{j\omega_eT_S}(z \cdot e^{j\omega_eT_S} - e^{-\beta}) + \dfrac{R_aT_S}{L}}$$

$$\approx \frac{\dfrac{T_S}{L}}{z(z - 1) + \dfrac{R_aT_S}{L}}$$

Stability of the modified load W_{LRA} depends on the factor R_aT_S/L. The maximum value of R_aT_S/L is 1; any larger value introduces instability. Thus, the maximum value of R_a is $L/T_S = 200$.

Question (7.2) Digital current control system of Question 7.1 has one-PWM-period feedback averaging. It is necessary to calculate the maximum stable value of the active resistance feedback R_a.

Answer (7.2) The feedback averaging introduces the pulse transfer function W_{FB} into the feedback path.

$$W_{FB}(z) = \frac{i^{FB}(z)}{i(z)} = \frac{z^2 + 2 \cdot z + 1}{4 \cdot z^2}$$

Modified pulse transfer function is given in

$$W_{LRA}(z) = \frac{\frac{T_S}{L}}{z \cdot e^{j\omega_e T_S}\left(z \cdot e^{j\omega_e T_S} - e^{-\beta}\right) + \frac{R_a T_S}{L} \cdot \frac{z^2 + 2z + 1}{4z^2}}$$

$$\approx \frac{\frac{T_S}{L} z^2}{z^3(z - 1) + \frac{R_a T_S}{4L} \cdot (z^2 + 2z + 1)}.$$

Characteristic polynomial $f(z)$ is given in

$$f(z) = z^4 - z^3 + \frac{R_a T_S}{4L} z^2 + \frac{R_a T_S}{2L} z + \frac{R_a T_S}{4L}.$$

Stability can be tested by using Matlab and performing the search by means of the scripts similar to those proposed in Chap. 6, Questions 6.1, 6.2, 6.3, 6.4, 6.5, 6.6, and 6.7. The roots of the polynomial $f(z)$ are stable until the factor $R_a T_S/4L$ reaches the value of 0.1705. Thus, the maximum value of the feedback gain is $R_a = 0.1705 \cdot 4L/T_S = 136$.

Question (7.3) Digital current control system of Question 7.1 has one-PWM-period feedback averaging, and it uses the advanced scheduling scheme of Chap. 6, which reduces the transport delays. It is necessary to calculate the maximum stable value of the active resistance feedback R_a.

Answer (7.3) With the advanced scheduling, the pulse transfer function of the load is given in (6.8).

$$W_{L1}(z) = \frac{i^e(z)}{u^e(z)}\bigg|_{e^e=0} = \frac{T_S}{L} \frac{1}{z \cdot e^{j\omega_e T_S} - e^{-\beta}}.$$

The feedback averaging introduces the pulse transfer function W_{FB} into the feedback path,

$$W_{FB}(z) = \frac{i^{FB}(z)}{i(z)} = \frac{z^2 + 2 \cdot z + 1}{4 \cdot z^2}$$

Thus, the modified pulse transfer function is given by

$$W_{\text{LRA}}(z) = \frac{W_{\text{LI}}(z)}{1 + R_a W_{\text{FB}}(z) W_{\text{LI}}(z)}$$

$$= \frac{\dfrac{T_S}{L} \dfrac{1}{z \cdot e^{j\omega_e T_S} - e^{-\beta}}}{1 + \dfrac{z^2 + 2 \cdot z + 1}{4 \cdot z^2} \cdot \dfrac{R_a T_S}{L} \dfrac{1}{z \cdot e^{j\omega_e T_S} - e^{-\beta}}}$$

With the angular frequency, ω_e is close to zero and with the factor $\exp(-\beta)$ close to 1,

$$W_{\text{LRA}}(z) = \frac{\dfrac{T_S}{L} z^2}{z^3 - z^2 + (z^2 + 2 \cdot z + 1) \cdot \dfrac{R_a T_S}{4L}} = \frac{T_S}{L} \cdot \frac{z^2}{z^3 + \left(-1 + \dfrac{R_a T_S}{4L}\right) z^2 + z \dfrac{R_a T_S}{2L} + \dfrac{R_a T_S}{4L}}$$

Stability can be tested by performing the search by means of Matlab scripts similar to those in Questions 6.1, 6.2, 6.3, 6.4, 6.5, 6.6, and 6.7. The roots of the polynomial $f(z)$ are stable until the factor $R_a T_S/4L$ reaches the value of 0.333. Thus, the maximum value of the feedback gain is $R_a = 0.333 \cdot 4L/T_S = 267$.

Chapter 8
Synchronization and Control

The grids for transmission and distribution of electric energy have an ever-increasing share of static power converters. They include the source-side converters, the bus converters, and the load-side converters. Typical source-side converters are the inverters that collect the electric energy from the wind power plants or solar power plants, convert the energy into a set of three-phase voltages and currents, and inject the active and reactive power into the three-phase ac grid. The bus power converters are used for connections between the grids of different voltage levels, and they can be either ac/ac, dc/dc, or ac/dc. In a way, the bus converters tend to replace the traditional line-frequency power transformers. The load-side power converters are used as the power interface between the grid and the load. They convert the grid voltages and adjust the load voltages to suit the needs of the electric power application. With the advent of local accumulation and considering the regeneration needs of electrical drives, most load-side converters have to be bidirectional, capable of supplying electric energy into the grid during brief intervals of time. Therefore, the basic functionality of all the grid-side converters is similar. When interfacing the ac grids, the grid-side power converter has to provide the voltages and inject the currents that are in synchronism with the grid voltages. Therefore, it is necessary to provide the means for detecting the frequency and the phase of the grid ac voltages. Most common device in use is the phase-locked loop (PLL), often used in radio circuits.

Dynamic properties of the grid-synchronization device have a significant impact on the response of the grid-side converter power to the grid transients. Over the past century, the ac grid resources, controls, and protections have been designed for the power system where all the sources are synchronous generators. With the present infrastructure, the power response of the grid-side converter to the grid transients should, ideally, resemble the response of a typical synchronous generator. An increased share of grid-side power converters with different dynamic properties could have an adverse effect on the grid operation and stability. Therefore, it is of interest to study the grid-synchronization devices and to find the way to make their dynamic properties closer to those of the synchronous generators.

© Springer International Publishing AG, part of Springer Nature 2018
S. N. Vukosavic, *Grid-Side Converters Control and Design*, Power Electronics
and Power Systems, https://doi.org/10.1007/978-3-319-73278-7_8

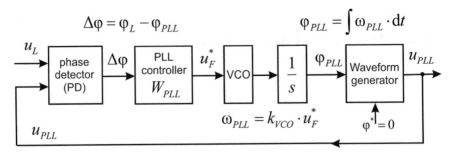

Fig. 8.1 Basic structure of the phase-locked loop (PLL)

8.1 Phase-Locked Loop

The phase-locked loop (PLL) is often used in radio circuits and other applications to detect the frequency and phase of the incoming signal and/or to generate the ac signal with the same frequency and with the desired phase shift with respect to the incoming signal. The basic structure of the PLL is given in Fig. 8.1.

In Fig. 8.1, it is assumed that the input signal u_L represents a sinusoidal signal of the frequency ω_L or that the sinusoidal signal represents a significant component of the input signal. The main objective of the structure in Fig. 8.1 is generation of the output signal u_{PLL}, a sinusoidal signal with the same frequency and phase as the input signal u_L. The input signal u_L and the output of the waveform generator u_{PLL} are brought to the phase detector PD, which generates the signal $\Delta\varphi$, proportional to the phase displacement between u_L and u_{PLL}. The next block in the chain is the PLL controller. It has the transfer function W_{PLL}, and it is most frequently designed as a PI controller. The PLL controller generates the voltage command u^*_F which is brought to the voltage-controlled oscillator (VCO). The VCO generates the output at the frequency which is proportional to the analogue input signal; therefore, $\omega_{PLL} = k_{VCO} \cdot u^*_F$. The phase of that signal is designated by φ_{PLL}, and it is obtained by passing the frequency ω_{PLL} through an integrator. In Fig. 8.1, the waveform generator creates an ac signal at the frequency ω_{PLL} and with the phase equal to $\varphi_{PLL} + \varphi^*$. The phase reference φ^* can be inserted in order to control and alter the phase shift of u_{PLL} with respect to the phase of the input signal u_L. In further considerations, it is assumed that the phase reference φ^* is equal to zero and that the circuit of Fig. 8.1 reaches the steady state when the phases of the signals u_L and u_{PLL} are equal.

Static and dynamic characteristics of the PLL circuit of Fig. 8.1 depend on the properties of the phase detector, on the gains used within W_{PLL}, and on the VCO gain k_{VCO}.

8.1.1 The Phase Detector with a Multiplier

The purpose of the phase detector is to provide the output signal which is proportional to the phase shift between the input signal u_L and the output of the waveform generator u_{PLL}. Assuming that the input signal is a sinusoidal waveform of the amplitude U_L, of the frequency ω_L, and with the phase-shift φ_L and that the waveform generator of Fig. 8.1 produces the signal u_{PLL} as a cosine function of the amplitude U_{PLL}, of the frequency ω_{PLL}, and with the phase-shift φ_{PLL}, the two signals can be written as

$$u_L(t) = U_L \cdot \sin(\omega_L t + \varphi_L),$$
$$u_{PLL}(t) = U_{PLL} \cdot \cos(\omega_{PLL} t + \varphi_{PLL}). \tag{8.1}$$

There are several ways to design the phase detection circuit. One of them relies on the analogue multiplier which generates the product $u_L \cdot u_{PLL}$ filters out the high-frequency component of the outcome and uses the average value, which is used as an indication of the phase difference. The product $u_L \cdot u_{PLL}$ is given in (8.2).

$$u_L(t) \cdot u_{PLL}(t) = U_L \cdot U_{PLL} \cdot \sin(\omega_L t + \varphi_L) \cdot \cos(\omega_{PLL} t + \varphi_{PLL}) =$$
$$= \frac{U_L \cdot U_{PLL}}{2}\left[\sin(\omega_L t + \varphi_L - \omega_{PLL} t - \varphi_{PLL}) + \sin(\omega_L t + \varphi_L + \omega_{PLL} t + \varphi_{PLL})\right]$$
$$\tag{8.2}$$

Assuming that the two signals are already locked ($\omega_L = \omega_{PLL}$) or that their frequencies are close, the signal $\sin(\omega_L t + \varphi_L + \omega_{PLL} + \varphi_{PLL})$ changes at frequency two times larger than the input frequency, and it has an average value of zero. The phase detector based on the multiplication of the two input signals has a low-pass filter which removes the signal at the frequency $2\omega_L$. The cutoff frequency of such filter is considerably larger than the desired bandwidth of the PLL. Therefore, such low-pass filter is disregarded both in Fig. 8.1 and in further considerations.

By removing the fast-changing ac component from (8.2), what remains is the signal $\sin(\omega_L t + \varphi_L - \omega_{PLL} - \varphi_{PLL})$. In cases where the two signals are already locked ($\omega_L = \omega_{PLL}$), the output of demodulator remains proportional to $\sin(\varphi_L - \varphi_{PLL})$. For small values of $\Delta\varphi = \varphi_L - \varphi_{PLL}$, detected signal is proportional to $\Delta\varphi$. Thus, the PLL controller W_{PLL} can change the VCO input voltage in the way that changes the PLL frequency and drives the output phase φ_{PLL} toward φ_L. In transient conditions where $\omega_L \neq \omega_{PLL}$, the phase detector provides the output signal which exhibits slow changes at the frequency $\Delta\omega = \omega_L - \omega_{PLL}$. In cases where the difference $\Delta\omega$ is sufficiently low, the slow changing signal passes through the low-pass elements of the closed-loop system of Fig. 8.1, it affects the VCO and brings the PLL frequency ω_{PLL} to ω_L, the PLL *locks* to the input signal, and the system enters the steady-state conditions. The largest value of $|\Delta\omega|$ that still permits the PLL to lock to the input signal depends on the gains of VCO and W_{PLL}.

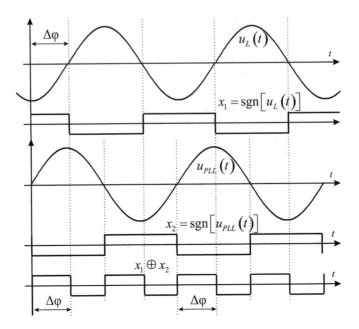

Fig. 8.2 Phase detection based on the exclusive OR function

8.1.2   Phase Detector with XOR Function

The phase detection based on the analogue multiplier provides the output which is not linear but a sine function of the phase error. At the same time, the analogue multiplier is a nonconventional part, difficult to implement, unless the input signals u_L and u_{PLL} are fast-sampled and brought into the digital signal processor, where all further processing is performed in digital, discrete-time domain. Another way of implementing the phase detection circuit is the use of the exclusive-OR logical function, illustrated in Fig. 8.2.

In Fig. 8.2, the logical signals x_1 and x_2 represent the sign of the input signal u_L and the signal u_{PLL}. The trace shown at the bottom of Fig. 8.2 represents the exclusive OR function of x_1 and x_2. The dwell time of the positive pulses of $x_1 \oplus x_2$ corresponds to the phase error. When the output of the exclusive-OR gate gets passed through the low-pass filter, the pulsating components are removed, leaving only the average value. From Fig. 8.2, the average value of $x_1 \oplus x_2$ is a linear function of the phase error $\Delta\varphi$.

8.1.3   The Phase Detector Based on the d-Axis Voltage

In grid-side converters, the phase-locked loop is used to synchronize the d-q frame to the line voltages. In symmetrical three-phase systems, the line voltage can be represented by the revolving vector with an amplitude that corresponds to the line-

Fig. 8.3 Synchronization of the *d-q* frame to the line voltage. The *q*-axis current controls the active power, while the *d*-axis current controls the reactive power injected from the grid-side converter into the grid

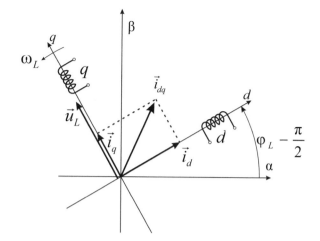

voltage amplitude, with the frequency that corresponds to the line frequency, and with the spatial orientation determined by the phase of the line voltages. If the *d-q* frame revolves in synchronism (i.e., with the same revolving speed) with the line-voltage vector, while at the same time the *q*-axis maintains the same spatial orientation as the line-voltage vector (i.e., the voltage vector and the *q*-axis remain collinear), then the *q*-axis current of the grid-side converter controls the active power, while the *d*-axis power controls the reactive power (Fig. 8.3).

The use of the synchronous *d-q* frame in grid-side converters can be advantageously used to implement a simple and effective phase detector.

In Fig. 8.3, it is assumed that the phase of the line voltage is obtained from the PLL and that the angle φ_{PLL} corresponds to the phase of the line voltage φ_L. In such case, the projection of the line-voltage vector (Fig. 8.3) on the *d*-axis of the synchronous frame is equal to zero. Thus, coordinate transformation of the line voltages into the *d-q* frame would result in $u_d = 0$. In cases where the line voltage vector drifts away from the *q*-axis in a positive, counterclockwise direction, projection of the line voltage on the *d*-axis becomes negative (Fig. 8.4).

The *d-q* frame of Fig. 8.4 is synchronized to the line voltage by the phase-locked loop, namely, the position of the *q*-axis is determined by the PLL output angle φ_{PLL}. Thus, if the line voltage advances by $\Delta\varphi = \varphi_L - \varphi_{\mathrm{PLL}}$, the line voltage vector would lead ahead of the *q*-axis by the phase error $\Delta\varphi$. Thus, once the line voltages are sampled and transformed into the *d-q* frame, the *d*-axis component of the line voltage assumes a nonzero value, proportional to $-\sin(\Delta\varphi)$. Therefore, the *d*-axis component of the line voltage obtained from a PLL-synchronized *d-q* frame can be used as an indicator of the phase error. In the prescribed manner, the phase detection is implemented without an additional effort (Fig. 8.5).

Fig. 8.4 The phase error $\Delta\varphi$ places the vector of the line voltage ahead of the q-axis. The d-q frame is synchronized by using the PLL angle φ_{PLL}. Due to the phase error, the line voltage projection on d-axis assumes a nonzero value proportional to the sine of the phase error

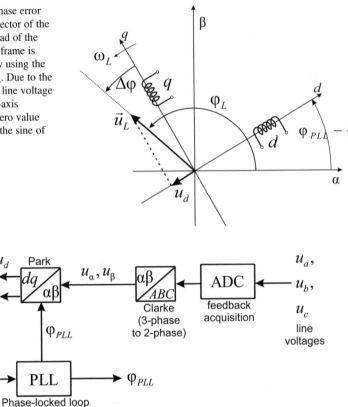

Fig. 8.5 The phase detector within the phase-locked loop can be replaced by relying on the d-axis component of the line voltage, obtained in the d-q frame that is synchronized by using the PLL angle φ_{PLL}. The line voltage projection on d-axis assumes a nonzero value proportional to the sine of the phase error

8.1.4   The Closed-Loop Operation of the PLL

It is of interest to obtain the closed-loop transfer function of the phase-locked loop of Fig. 8.1. For this reason, it is necessary to analyze the elements of the closed-loop system and to derive the corresponding transfer functions. The phase detector receives the signals u_L and u_{PLL}, and it derives the output signal which is proportional to the phase difference $\Delta\varphi = \varphi_L - \varphi_{\text{PLL}}$, namely, it outputs the signal $k_{\text{PD}} \cdot \Delta\varphi$. Assuming that the PLL controller W_{PLL} has the proportional gain k_{P} and the integral gain k_{i}, the output of the PLL controller is given by

$$u_F^*(s) = k_{\text{PD}} \cdot \left(k_{\text{P}} + \frac{k_{\text{i}}}{s}\right) \cdot \Delta\varphi(s) \tag{8.3}$$

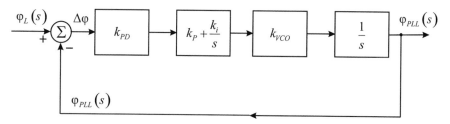

Fig. 8.6 Simplified block diagram of the PLL

The signal $u^*{}_F$ is brought to the voltage-controlled oscillator, which provides the sinusoidal output at the frequency $\omega_{PLL} = k_{VCO} \cdot u^*{}_F$. The waveform generator of Fig. 8.1 provides the signal u_{PLL}. The phase φ_{PLL} of the signal u_{PLL} is defined by

$$\varphi_{PLL}(s) = \frac{k_{VCO}}{s} \cdot u_F^*(s) = \frac{k_{VCO}}{s} \cdot k_{PD} \cdot \left(k_P + \frac{k_i}{s} \right) \cdot \Delta\varphi(s) \qquad (8.4)$$

The phase of the input signal u_L can be treated as the input to the PLL loop, while the phase φ_{PLL} of the output signal u_{PLL} can be treated as the output. Thus, the block diagram of Fig. 8.1 can be simplified and shown in Fig. 8.6.

The system of Fig. 8.6 has the open-loop gain equal to

$$W_{OG}(s) = \frac{\varphi_{PLL}(s)}{\Delta\varphi(s)} = \frac{s \cdot k_P k_{PD} k_{VCO} + k_i k_{PD} k_{VCO}}{s^2}. \qquad (8.5)$$

The closed-loop transfer function $W_{CL}(s)$ is equal to $\varphi_{PLL}(s)/\varphi_L(s)$, and it is obtained from $W_{OG}(s)/(1 + W_{OG}(s))$:

$$W_{CL}(s) = \frac{\varphi_{PLL}(s)}{\varphi_L(s)} = \frac{s \cdot k_P k_{PD} k_{VCO} + k_i k_{PD} k_{VCO}}{s^2 + s \cdot k_P k_{PD} k_{VCO} + k_i k_{PD} k_{VCO}}. \qquad (8.6)$$

Assuming that the input phase exhibits a step change, it is represented by $\varphi_L(s) = J/s$. It this case, the output phase is represented as

$$\varphi_{PLL}(s) = \frac{J}{s} \cdot \frac{s \cdot k_P k_{PD} k_{VCO} + k_i k_{PD} k_{VCO}}{s^2 + s \cdot k_P k_{PD} k_{VCO} + k_i k_{PD} k_{VCO}}. \qquad (8.7)$$

In steady state, the output position φ_{PLL} reaches the input step $\varphi_L = J$ without an error:

$$\begin{aligned}
\varphi_{PLL}(\infty) &= \lim_{s \to 0} \left(s \cdot \varphi_{PLL}(s) \right) \\
&= \left(s \cdot \frac{J}{s} \cdot \frac{s \cdot k_P k_{PD} k_{VCO} + k_i k_{PD} k_{VCO}}{s^2 + s \cdot k_P k_{PD} k_{VCO} + k_i k_{PD} k_{VCO}} \right)_{s=0} = J.
\end{aligned} \qquad (8.8)$$

Dynamic response of the PLL is defined by the closed-loop poles. With the assumption that the PLL has one pair of conjugate complex poles, the characteristic polynomial in the denominator of (8.6) can be expressed as $f(s) = s^2 + 2 \cdot \xi \cdot \omega_n \cdot s + \omega_n^2$, where ω_n is the natural frequency of the closed-loop poles and ξ is the damping of the poles. With $\xi < 0.5$, the step response has an oscillatory character, while $\xi > 1$ provides an aperiodic response. Thus, the use of the PLL controller with $k_P > 0$ is required in order to suppress undamped oscillations that would appear otherwise. The natural frequency determines the closed-loop bandwidth of the phase-locked loop. With $\xi = 0.707$, the closed-loop bandwidth $f_{BW}(-3\ \text{dB})$ corresponds to the natural frequency of the closed-loop poles. With the natural frequency determined by $\omega_n^2 = k_i \cdot k_{PD} \cdot k_{VCO}$, the closed-loop bandwidth of the loop is defined by the integral gain k_i. In most cases, the phase-locked loops that are used in grid-side inverters have the closed-loop bandwidth from 1 Hz up to 10 Hz.

8.2 Dynamic Response of Grid-Side Converters

The number of power sources and loads that are connected to the ac grid by means of electronically controlled grid-side converters keeps increasing in number and also in their share of the total installed power. Therefore, dynamic response of grid-side converters becomes ever more important. The ac grid and its control and protection mechanisms have been designed for the use in conjunction with traditional three-phase, wound-field synchronous generators. For this reason, all the while the ac grids are maintaining the present control concepts, it is of interest to introduce the control means into the grid-side converters that would bring their dynamic properties closer to those of the synchronous generators.

Most grid-side converters use PLL synchronization, and they incorporate the voltage, current, and power controllers capable of maintaining the desired active power in the presence of all the usual grid-side disturbances. Typical grid-side disturbances include the changes in the line voltage amplitude and the changes in the line voltage phase and frequency. Thus, it is of interest to study the response of the conventional synchronous machines and traditional loads to the voltage-amplitude and voltage-phase disturbances that originate from the grid. It is also of interest to study the response of PLL-synchronized, electronically controlled grid-side inverters to the same set of disturbances and to devise the control means to bring the later responses closer to those of conventional sources and loads.

8.2.1 Dynamic Response of Conventional Generators

Conventional synchronous machines comprise the three-phase stator winding, the excitation winding on the rotor, and the damping winding on the rotor. In large turbo generators, dynamic processes within the damping winding decay in $0.02 \div 0.05$ s, and they determine the *subtransient* phenomena of the generator transients. Dynamic processes within the excitation winding decay in $0.4 \div 0.9$ s, and they define the *transient* phenomena. Both processes have considerable impact on the generator capability to supply the fault currents in the case of the short circuit, and they largely affect the change and the envelope of the short circuit currents. For this reason, subtransient and transient phenomena have the key role in designing and tuning the ac grid protection mechanisms. The capability of grid-side converters to emulate the short circuit behavior of synchronous generators is limited by the current capacity of semiconductor power switches, and it is not in focus of the discussion within this section.

In the course of the subsequent analysis, the processes in the damper windings are neglected, and it is assumed that the excitation current I_{EXC} is constant. Thus, the rotor flux that encircles the stator winding is also constant and equal to $\Psi_{Rm} = L_m \cdot I_{EXC}$. In steady state, the rotor speed Ω_m is equal to the synchronous speed Ω_e. The frequency of grid-connected generator stator currents is equal to the line-frequency ω_L. In synchronous generator with p pole pairs, the synchronous speed is equal to $\Omega_e = \omega_L/p$, and it determines the speed of rotation of the stator magnetic field. In steady state, $\Omega_m = \Omega_e = \omega_L/p$. Large turbo generators are mostly two-pole machines with $p = 1$, $\Omega_e = \omega_L$, and $\Omega_m = \omega_m$, which is the assumption considered in all further discussions. The objective of the subsequent developments is deriving of simple-to-use expressions that reflect the dynamic response of the synchronous generator active power to the changes of the grid voltage amplitude and phase.

Synchronous generators are often modeled in synchronously rotating d-q coordinate frame where d-axis is collinear with the excitation flux. Therefore, d-q frame revolves at the rotor speed Ω_m. For isotropic synchronous machine (the ones where the magnetic circuit maintains the same magnetic resistance in all directions, and where $L_s = L_d = L_q$), the voltage balance equations of the stator windings in d-q frame are

$$
\begin{aligned}
u_d &= R_s i_d + L_s \frac{di_d}{dt} - \omega_m L_s i_q, \\
u_q &= R_s i_q + L_s \frac{di_q}{dt} + \omega_m L_s i_d + \omega_m \Psi_{Rm},
\end{aligned}
\tag{8.9}
$$

where L_s is the synchronous inductance of the stator winding, R_s is the resistance of the stator windings, ω_m is the revolving speed of the rotor, u_d and u_q are the projections of the stator voltage vector on the axes of the rotor-locked d-q frame, and i_d and i_q are the projections of the stator currents on the same axes.

The considered synchronous generator is connected to the ac grid, and the stator voltages u_d and u_q represent the grid voltages at the connection point. The stator currents i_d and i_q are intended as the currents which circulate from the grid into the stator winding. The processes in the damping, excitation, and the stator windings are relatively fast, and they quickly decay into the steady state. Therefore, the voltage balance equation (8.9) can be simplified by removing the derivatives of the stator currents:

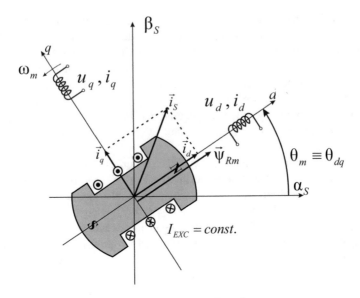

Fig. 8.7 Synchronous machine represented in d-q coordinate frame

$$
\begin{aligned}
u_d &= R_s i_d - \omega_{\mathrm{m}} L_s i_q \\
u_q &= R_s i_q + \omega_{\mathrm{m}} L_s i_d + \omega_{\mathrm{m}} \Psi_{\mathrm{Rm}}.
\end{aligned}
\tag{8.10}
$$

The stator currents and voltages can be represented by complex numbers $\underline{u}_s = u_d + j u_q$ and $\underline{i}_s = i_d + j i_q$. By multiplying the second equation in (8.10) by j and adding the result to the first equation, one obtains

$$
\underline{u}_s = R_s \underline{i}_s + j \omega_{\mathrm{m}} L_s \underline{i}_s + j \omega_{\mathrm{m}} \Psi_{\mathrm{Rm}}.
\tag{8.11}
$$

The voltage balance equation (8.11) can be represented by the phasor diagram shown in Fig. 8.8, where the electromotive force $\underline{E}_0 = j\omega_{\mathrm{m}} \Psi_{\mathrm{Rm}}$ is collinear with the q-axis while the line voltage $\underline{u}_s$ lags by the angle $\delta > 0$.

The active and reactive power injected into the grid can be obtained from the stator voltages and currents. In order to obtain the stator currents, it is necessary to solve eq. (8.11). Since the stator voltage lags with respect to $\underline{E}_0$ by δ, the stator voltage components can be expressed as $u_d = U_S \sin(\delta)$ and $u_q = U_S \cos(\delta)$, where U_S is the amplitude of the line voltage. Therefore,

$$
\begin{aligned}
\underline{u}_S &= U_S \sin \delta + j U_S \cos \delta, \\
\underline{i}_S &= \frac{\underline{u}_S - \underline{E}_0}{j L_s \omega_{\mathrm{m}}} = \frac{U_S \sin \delta + j U_S \cos \delta - j \omega_{\mathrm{m}} \Psi_{\mathrm{Rm}}}{j L_s \omega_{\mathrm{m}}} = \\
&= \frac{U_S \cos \delta - \omega_{\mathrm{m}} \Psi_{\mathrm{Rm}}}{L_s \omega_{\mathrm{m}}} - j \frac{U_S \sin \delta}{L_s \omega_{\mathrm{m}}}
\end{aligned}
\tag{8.12}
$$

The power injected from the synchronous generator into the grid is obtained as

Fig. 8.8 Phasor diagram of the synchronous machine

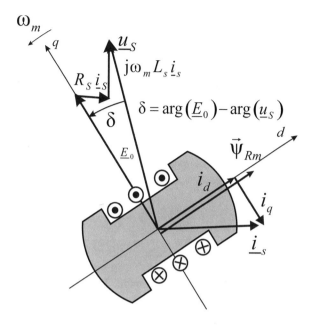

$$\underline{S}_G = -3 \cdot \underline{u}_S \underline{i}_S^* = P_G + jQ_G =$$
$$= 3 \cdot \frac{U_S(\omega_m\Psi_{Rm})}{L_S\omega_m} \sin(\delta) + j \cdot 3 \cdot \frac{U_S(\omega_m\Psi_{Rm}\cos(\delta) - U_S)}{L_S\omega_m}. \qquad (8.13)$$

It is of interest to analyze the change of the power P_G in cases where the grid voltage exhibits changes in amplitude and phase. Assuming that the voltage amplitude increases from U_{S0} to $U_{S0} + \Delta U$, while the phase shift of the grid voltage relative to the rotor remains constant, the power supplied from the generator changes to

$$P_G = 3 \cdot \frac{(U_{S0} + \Delta U)(\omega_m\Psi_{Rm})}{L_S\omega_m} \sin(\delta),$$
$$\Rightarrow \frac{dP_G}{dU_S} = 3 \cdot \frac{(\omega_m\Psi_{Rm})}{L_S\omega_m} \sin(\delta) = \frac{P_G}{U_S}. \qquad (8.14)$$

In cases where the phase shift of the grid voltages changes from the initial φ_{L0} to $\varphi_{L0} + \Delta\varphi$, while, at the same time, the rotor speed remains unaltered, the angle δ would change from the initial δ_0 to $\delta_0 - \Delta\varphi$, and this change would affect the power P_G supplied to the grid:

$$P_G = 3 \cdot \frac{U_S(\omega_m\Psi_{Rm})}{L_S\omega_m} \sin(\delta_0 - \Delta\varphi) = P_m \sin(\delta_0 - \Delta\varphi),$$
$$\Rightarrow \frac{dP_G}{d(\Delta\varphi)} = -3 \cdot \frac{U_S(\omega_m\Psi_{Rm})}{L_S\omega_m} \cos(\delta_0 - \Delta\varphi) = -P_m \cos(\delta_0 - \Delta\varphi). \qquad (8.15)$$

From (8.14), it is visible that the generator power increases with the rise and decreases with the fall of the grid voltage amplitude. At the same time, expression (8.15) provides the evidence that the generator power decreases with the advance of the grid-voltage phase shift.

Expression (8.15) is obtained under assumption that the rotor speed remains unaltered during the phase-shift transient. The change of the angle δ to $\delta_0 - \Delta\varphi$ changes the generator power and the electromagnetic torque. In turn, the torque change affects the rotor speed, thus invalidating the original assumption, which could hold only in cases with infinitely large rotor inertia. In order to calculate the power transient accurately, it is necessary to take into account the differential equation which describes the changes of the rotor speed:

$$J \cdot \frac{d\omega_m}{dt} = T_T - T_{em} = T_T - \frac{P_G}{\omega_m} \approx T_{em0} - \frac{P_G}{\omega_L}. \tag{8.16}$$

In (8.16), J is the equivalent inertia of the rotor, the steam (water) turbine, and the relevant shafts and couplings. The moving torque T_T is provided from the turbine. For the purposes of this discussion, it can be considered constant and equal to the steady-state value of the electromagnetic torque $T_{em} = T_{em0} = P_{G0}/\omega_m$. In (8.16), the friction and other minor motion resistances are neglected, which is a quite reasonable assumption with large synchronous generators.

From (8.15), the power P_G can be expressed as $P_m \sin(\delta)$, where P_m is the maximum power, obtained with $\delta = \pi/2$. In Fig. 8.8, position of d-axis is equal to θ_m, and the angle δ is equal to

$$\delta = \theta_m + \frac{\pi}{2} - \varphi_L \tag{8.17}$$

It is of interest to study the changes ΔP_G of the generator power caused by small changes of the grid-voltage phase $\Delta\varphi_L$. With corresponding change of the rotor position $\Delta\theta_m$, the angle δ changes to $\delta = \delta_0 + \Delta\delta$. From (8.17), $\Delta\delta = \Delta\theta_m - \Delta\varphi_L$. For small disturbances, $\delta_0 \gg \Delta\delta$, and the change of the generator power can be expressed as

$$\Delta P_G = P_m \cos(\delta_0) \cdot \Delta\delta = P_m \cos(\delta_0) \cdot (\Delta\theta_m - \Delta\varphi_L). \tag{8.18}$$

From (8.16),

$$\omega_L \cdot J \cdot \frac{d^2}{dt^2}(\Delta\theta_m) = -\Delta P_G = -P_m \cos(\delta_0) \cdot (\Delta\theta_m - \Delta\varphi_L). \tag{8.19}$$

By means of the Laplace transformation, differential equation (8.19) can be converted into s-domain and solved for $\Delta\theta_m$:

$$\omega_L \cdot J \cdot s^2 \Delta\theta_m(s) = P_m \cos(\delta_0) \cdot \Delta\varphi_L(s) - P_m \cos(\delta_0) \cdot \Delta\theta_m(s),$$
$$\Delta\theta_m(s) = \frac{P_m \cos(\delta_0)}{\omega_L \cdot J \cdot s^2 + P_m \cos(\delta_0)} \cdot \Delta\varphi_L(s). \qquad (8.20)$$

Introducing the result (8.20) into (8.18), one obtains the response of the synchronous-generator power ΔP_G to the changes of the grid-voltage phase $\Delta\varphi_L$:

$$\begin{aligned}
\Delta P_G &= P_m \cos(\delta_0) \cdot [\Delta\theta_m(s) - \Delta\varphi_L(s)] = \\
&= P_m \cos(\delta_0) \cdot \left[\frac{P_m \cos(\delta_0)}{\omega_L \cdot J \cdot s^2 + P_m \cos(\delta_0)} - 1 \right] \cdot \Delta\varphi_L(s) = \\
&= \frac{-\omega_L \cdot J \cdot s^2 \cdot P_m \cos(\delta_0)}{\omega_L \cdot J \cdot s^2 + P_m \cos(\delta_0)} \cdot \Delta\varphi_L(s).
\end{aligned} \qquad (8.21)$$

The frequency that characterizes the response of the generator power to the grid disturbances is obtained from the characteristic polynomial in the denominator of (8.21):

$$\begin{aligned}
f(s) &= \omega_L \cdot J \cdot s^2 + P_m \cos(\delta_0), \\
p_{1/2} &= \pm j \cdot \sqrt{\frac{P_m \cos(\delta_0)}{\omega_L \cdot J}}.
\end{aligned} \qquad (8.22)$$

Inertia constant $H = (J \, \omega_L^2/2)/S_{nom}$ of large turbo generators ranges from 4 up to 8 MW · s/MVA. Assuming that the factor $P_m \cos(\delta_0)$ is close to the rated power S_{nom}, the frequency of the poles $p_{1/2}$ of (8.22) ranges from 0.7 Hz up to 1 Hz.

The response of ΔP_G to the phase disturbance is determined by the transfer function $-s^2/(s^2 + p^2)$, while the response to the frequency disturbance depends on $-s/(s^2 + p^2)$. The characteristic polynomial of (8.22) does not imply any damping. Namely, the response of (8.21) is poorly damped, and the grid transients would cause oscillations that would not decay. In practical synchronous machines, the response of ΔP_G to the grid disturbances are suppressed by the damper winding, the rotor winding that resembles the short circuited cage winding of induction machines.

8.2.2 The Impact of the Damper Winding

In addition to the three-phase stator winding and the excitation winding, the synchronous generators also have the damping winding on the rotor, introduced with the aim to suppress the oscillations introduced by the conjugate complex poles of (8.22). The damper winding resembles the squirrel cage in induction machine, and it generates an additional component of the electromagnetic torque, proportional to the slip, namely, to the relative speed of the stator voltages and currents with respect to the rotor. With damper winding, Eq. (8.16) changes into (8.23). The first derivative of $\varphi_L - \theta_m$ represents the slip, while the coefficient k_{DW} depends on the characteristics of the damper winding.

$$J \cdot \frac{d\omega_m}{dt} = T_{em0} - \frac{P_m}{\omega_L} \sin(\delta) + \frac{k_{DW}}{\omega_L} \frac{d(\varphi_L - \theta_m)}{dt}. \tag{8.23}$$

By considering the small, incremental changes of all the angles, and by transforming Eq. (8.23) in s-domain, it is possible to provide the expression for $\Delta\theta_m$ in terms of $\Delta\varphi_L$:

$$\omega_L \cdot J \cdot s^2 \Delta\theta_m(s) = P_m \cos(\delta_0) \cdot \Delta\varphi_L(s)$$
$$-P_m \cos(\delta_0) \cdot \Delta\theta_m(s) + k_{DW} \cdot s \cdot \Delta\varphi_L(s) - k_{DW} \cdot s \cdot \Delta\theta_m(s). \tag{8.24}$$

$$\Delta\theta_m(s) = \frac{P_m \cos(\delta_0) + k_{DW} \cdot s}{\omega_L \cdot J \cdot s^2 + k_{DW} \cdot s + P_m \cos(\delta_0)} \cdot \Delta\varphi_L(s). \tag{8.25}$$

Characteristic polynomial of (8.25) and the closed-loop poles are given in (8.26):

$$f(s) = s^2 + \frac{k_{DW}}{\omega_L \cdot J} \cdot s + \frac{P_m \cos(\delta_0)}{\omega_L \cdot J},$$
$$p_{1/2} = -\frac{k_{DW}}{2 \cdot \omega_L \cdot J} \pm j \cdot \sqrt{\frac{P_m \cos(\delta_0)}{\omega_L \cdot J} - \left(\frac{k_{DW}}{2 \cdot \omega_L \cdot J}\right)^2}. \tag{8.26}$$

Compared to (8.22), the poles of (8.26) have a nonzero real part, namely, the oscillations will decay with the time constant $2\omega_L J/k_{DW}$. Once the steady state is reached again, the slip gets equal to zero, the damper winding currents decay to zero, as well as the damping torque component produced by the damper winding.

The change of the angle δ is obtained as

$$\Delta\delta(s) = \Delta\theta_m(s) - \Delta\varphi_L(s) =$$
$$= \frac{-\omega_L \cdot J \cdot s^2}{\omega_L \cdot J \cdot s^2 + k_{DW} \cdot s + P_m \cos(\delta_0)} \cdot \Delta\varphi_L(s). \tag{8.27}$$

From (8.23), the electromagnetic torque of the generator has the component proportional to P_m, defined by the rotor excitation and the stator currents, and the component proportional to k_{DW}, which is defined by the currents in the damper winding. Thus, the change of the power delivered from the generator to the grid is

$$\Delta P_G = [P_m \cos(\delta_0) + s \cdot k_{DW}] \cdot \Delta\delta(s) =$$
$$= \frac{(-\omega_L \cdot J \cdot s^2) \cdot [P_m \cos(\delta_0) + s \cdot k_{DW}]}{\omega_L \cdot J \cdot s^2 + k_{DW} \cdot s + P_m \cos(\delta_0)} \cdot \Delta\varphi_L(s). \tag{8.28}$$

8.2.3 Dynamic Response of the PLL-Driven Converter

The power obtained from the wind power stations and the solar plants is processed through the source-side inverters, the three-phase inverters that inject controlled,

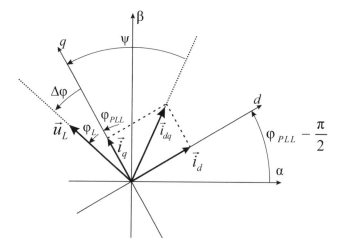

Fig. 8.9 Stator voltages and currents in PLL-synchronized d-q frame

line-frequency ac currents into the grid. The inverters receive the dc power that comes from the generator-side inverter (in the case of the win power station), or from the solar cells and their dc/dc converters. The dc power is converted in the power of the three-phase system of line-frequency voltages and currents.

With an increased use of electronically controlled sources, it is desirable to obtain their transient response as close to the one of traditional synchronous generators. In this way, the grid stability will not be affected by introduction of the electronically controlled sources. Namely, the transient response of grid-side inverters output power to changes in the grid frequency and phase should correspond to the response obtained with conventional synchronous generator. In most cases, the ac currents are controlled by the digital current controller located in the synchronous d-q frame, shown in Fig. 8.9.

The position of the q-axis is set to φ_{PLL}, where the angle φ_{PLL} is obtained from the phase-locked loop. In steady state, the angle φ_{PLL} is equal to the phase of the line voltages φ_L. Thus, the angle $\Delta\varphi$ between the line voltage and the q-axis drops to zero in the steady-state conditions. Since the line voltage is aligned with the q-axis, the current i_q determines the active power, while the current i_d determines the reactive power injected into the grid. The angle Ψ between the line voltage and the current i_{dq} depends on the ratio between the active and reactive power. When the reactive power is equal to zero, the angle Ψ is equal to zero as well. Notice in Fig. 8.9 that the reference direction of the current i_{dq} is from the grid-side inverter into the grid. Namely, positive current i_q contributes to positive active power injected into the grid, while positive current i_d contributes to positive reactive power.

When the frequency and the phase of the line voltages make a transient change, the angle φ_{PLL} falls behind φ_L, and the phase-locked loop enters the transient state defined by the transfer function (8.6). When the transient processes within the PLL settle down, the angle φ_{PLL} reaches φ_L. In the meantime, the angle $\Delta\varphi$ exhibits a transient change defined by

$$
\Delta\varphi(s) = \varphi_L(s) - \varphi_{PLL}(s) = \left[1 - \frac{s \cdot k_p k_{PD} k_{VCO} + k_i k_{PD} k_{VCO}}{s^2 + s \cdot k_p k_{PD} k_{VCO} + k_i k_{PD} k_{VCO}}\right] \cdot \varphi_L(s)
$$
$$
= \left[\frac{s^2}{s^2 + s \cdot k_p k_{PD} k_{VCO} + k_i k_{PD} k_{VCO}}\right] \cdot \varphi_L(s).
$$

(8.29)

From Fig. 8.9, the active power depends on the angle $\Delta\varphi + \Psi$ between the voltage u_L and the current i_{dq}:

$$
P_G = |\vec{u}_L| \cdot |\vec{i}_{dq}| \cdot \cos(\Delta\varphi + \psi) = U \cdot I \cdot \cos(\Delta\varphi + \psi).
$$

(8.30)

For small values of $\Delta\varphi$,

$$
\Delta P_G = -U \cdot I \cdot \sin(\psi) \cdot \Delta\varphi.
$$

(8.31)

By introducing (8.29) into (8.31), the power transients of the grid-side inverter are

$$
\Delta P_G(s) = -U \cdot I \cdot \sin(\psi) \cdot \Delta\varphi(s) =
$$
$$
= -\frac{U \cdot I \cdot \sin(\psi) \cdot s^2}{s^2 + s \cdot k_p k_{PD} k_{VCO} + k_i k_{PD} k_{VCO}} \cdot \varphi_L(s).
$$

(8.32)

The power transient obtained from the PLL-synchronized grid-side inverter (8.32) resembles the power transient obtained from the synchronous generator (8.28). Neglecting the power change obtained from the damper wining in (8.28), proportional to k_{DW}, the expressions (8.28) and (8.32) have the same structure.

$$
\Delta P_G = [P_m \cos(\delta_0) + s \cdot k_{DW}] \cdot \Delta\delta(s) =
$$
$$
= \frac{(-\omega_L \cdot J \cdot s^2) \cdot [P_m \cos(\delta_0) + s \cdot k_{DW}]}{\omega_L \cdot J \cdot s^2 + k_{DW} \cdot s + P_m \cos(\delta_0)} \cdot \Delta\varphi_L(s).
$$

(8.33)

If the factor $k_i k_{PD} k_{VCO}$ of the grid-side inverter is set to $P_m\cos(\delta_0)/\omega_L/J$, while the factor $k_p k_{PD} k_{VCO}$ is set to $k_{DW}/\omega_L/J$, the conjugate complex poles of denominators in (8.33) and (8.32) will be the same. If, at the same time, the factor $U \cdot I \cdot \sin(\Psi)$ assumes the value of $P_m\cos(\delta_0)$, the power transient of the grid-side inverter will be the same as the power transient obtained from the synchronous machine.

The factor $\cos(\delta_0)$ is strictly positive. It reaches zero only in cases where the synchronous machine operates with the maximum power, with $\delta_0 = \pi/2$. This extreme condition does not happen in normal operation. At the same time, the factor $\sin(\Psi)$ changes the sign power factor from leading to lagging. If the grid-side converter operates with unity power factor, where the reactive power is equal to zero, the current vector in Fig. 8.9 is aligned with the voltage vector, and the factor $\sin(\Psi)$ gets equal to zero. Therefore, dynamic response of the PLL-driven grid-side power converter is fundamentally different from the one of synchronous machines.

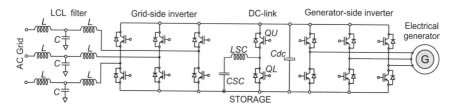

Fig. 8.10 Storage module within the wind power converter setup

With unity power factor, the grid-side inverters do not produce any power transients. Moreover, when the reactive power of grid-side converters is negative, their power transients change the sign and act against the power transients introduced by synchronous machines. In turn, this jeopardizes the stability in power grids with a large share of electronically controlled grid-side power converters.

In order to correct this undesired transient behavior of PLL-driven grid-side power converters, it is necessary to introduce some corrective control actions that would bring their transient power response closer to the ones of synchronous generators. One of the ways consists in introducing an additional component into the current reference for the q-axis. Notice in Fig. 8.6 that the angular difference $\Delta\varphi$ is one of the internal variables of the phase-locked loop. The transient response can be improved by adding the correction $k_{\text{corr1}}\Delta\varphi$ to the q-axis current reference. Desired power transients could also require a hardware modification of the power converter setup used in wind and solar power plants.

In the wind power converter setup of Fig. 8.10, the power obtained from the generator gets supplied into the dc-link; it is passed to the grid-side inverter and injected into the ac grid. The dc-link capacitor C_{dc} is not capable of storing or supplying a meaningful amount of energy. For this reason, and neglecting the power losses, the power injected into the ac grid has to be equal to the power obtained from the electrical generator. Namely, there is no possibility to supply additional power transients that are required to make the grid-side inverter resemble a synchronous machine. While synchronous machine has the rotor inertia as a readily available accumulator of the kinetic energy, the grid-side inverter with the dc-bus capacitor C_{dc} does not have such capability. Therefore, it is necessary to introduce a large capacitor, C_{SC} in Fig. 8.10, that would be capable of storing and regenerating the amounts of energy that are required to inject the desired power transients. The super-capacitor C_{SC} is controlled by two additional power switches (QU and QL) that control the current in the series inductance L_{SC}. Whenever the grid-side power gets different from the generator-side power, the current in the inductance L_{SC} is determined so as to maintain the dc-bus voltage unaffected. Thus, the excess energy is stored in C_{SC}, while the energy deficit is compensated by regenerating the energy from C_{SC}.

8.2.4 Emulation of Synchronous Machines

In addition to the transient response of the output power, conventional synchronous machine also exhibits other important transient features. They include the transient changes of the reactive power, the use of the rotor excitation to control the reactive power and the voltage conditions, and the change of the output current in the case of the short-circuit faults, the ground faults, and other fault conditions.

Conventional synchronous machine supplies the reactive power that depends on the difference between the grid voltage and the electromotive force which is proportional to the rotor flux and, thus, the excitation current. As the grid voltage increases, the reactive power injected into the grid reduces. In cases where the grid voltage exceeds the electromotive force, reactive power becomes negative, and the grid-side inverter starts absorbing the reactive power from the grid. In grid-side inverters with current control, with the voltages and currents shown in Fig. 8.9, reactive power is defined by the current references, and it does not decrease with an increase of the grid voltage. Quite on an opposite, in cases where i_d is constant and the grid voltage increases, the reactive power injected into the grid increases at the same time. This behavior is fundamentally different, and it could disturb the voltage-control mechanisms in ac grid.

In conventional synchronous machines, the amplitude of the grid ac voltages is maintained by adjusting the excitation current in the rotor circuit. Namely, in cases where the ac voltages are low, and there is a need to increase their amplitude, the outcome of the voltage-control mechanisms is an increase of the excitation current. In turn, there is an increase in the amplitude of the rotor flux and an increase in the induced electromotive force. As a consequence, the reactive power injected from the synchronous machine into the grid increases, contributing to a rise of the ac voltages. The grid-side inverter does not comprise an equivalent of the excitation current, but it can change the injected reactive power by altering the reference current i_d (Fig. 8.9). In practical applications, the grid operator runs the communication channels that can send the commands and references to the grid-side inverters that are distributed over the network. In cases where the voltage is low, and the need emerges to increase the reactive power, the operator send global or individual commands to the grid-side inverters to increase their i_d currents, increasing in this way the injection of reactive power. Such a mechanism could alleviate the voltage-control problems, but it does not have the same static and dynamic characteristics as those obtained with conventional synchronous machines.

The most pronounced difference between synchronous machines and grid-side inverters is observed in cases of the short circuit faults. The grid-side inverters that operate with an inner digital current controller maintain the output current at the prescribed reference value. When a short-circuit voltage appears in an ac grid, the voltage drops down, but the grid-side inverter continues to supply the output current which remains equal to the reference. In such cases, the reference can be increased intentionally, but it has to remain within the current rating of the semiconductor power switches. When the same event takes place with a synchronous machine, the short-circuit current that is supplied into the grid from the synchronous machines is considerably larger, and its amplitude changes during the subtransient, transient, and the steady-state phase of the short circuit fault. The first, subtransient period is defined by the electrical time constants of the damper winding, and it could take 20–50 ms to decay. The second, transient period is defined by the electrical time constants of the excitation winding, and it could take up to 1000 ms. When both the damping and the excitation windings enter the steady state, the fault current settles to the steady-state value. In subtransient period, the short-circuit current could exceed the rated current by and order of magnitude. The transient current is lower, while the steady-state fault current gets considerably lower.

In ac grids, it is of uttermost importance to detect the location of the short-circuit fault and to isolate the fault spot by opening the appropriate circuit breakers. It this way, most of the grid gets isolated from the fault spot, and the majority of the loads could be supplied without the power outages. In order to minimize the sage of the voltage, caused by the short-circuit fault, protection mechanisms have to be rather quick. At the same time, they have to identify the fault location precisely, in order to isolate this spot by opening the closest circuit breakers. Many protection mechanisms and fault-spot-location detection algorithms rely on the dynamic properties of conventional synchronous generators. For this reason, an ever-growing number of grid-side power converters change the fault current waveforms and jeopardize the traditional protection mechanism.

In order to maintain the traditional ac grid controls and protections, the newly emerging source-side inverters have to mimic the synchronous generator as close as possible. There are basically two ways of performing this task, the current-control way and the voltage-control way.

In grid-side inverters with an internal digital current controller, the output current is compelled to track the current references. In parallel with other control tasks, it is necessary to perform the real-time simulation of the synchronous generator. Namely, it is necessary to run the mathematical model of the pretended synchronous machine with both the excitation and the damper winding. Such model is supplied by the measured grid-side voltages, and it outputs the stator current of the would-be synchronous machine. At this point, it is sufficient to adopt the calculated currents as the current references of the grid-side inverter, and the inverter will inject the same ac currents as if it were the synchronous generator.

In grid-side inverters which do not have an internal current controller, the pulse width modulation sets the voltages that are supplied to the grid through the *LCL* filter. In very much the same way as with current-controlled inverters, it is necessary to run the mathematical model of the pretended synchronous machine with both the excitation and the damper winding. The model relies on the measured grid-side voltages, and it calculates the internal variables of the simulated synchronous machine, including the relevant flux linkages, currents, and electromotive forces. Namely, the model calculates the transient response that would be obtained from the synchronous machine if such machine were connected to the grid. From the real-time model, it is also possible to calculate the voltage that would be supplied to the input of the *LCL* filter from the simulated synchronous machine. At this point, the simulated voltage can be used as the modulation signal of the grid-side inverter. As a consequence, the inverter will inject the same ac currents as if it were the synchronous generator.

The implementation details of the voltage-mode method and the current-mode method are currently in the phase of development, and they are not discussed in this book. The hardware requirements for the proper implementation of the synchronous machine emulator include an increased current rating of the power semiconductor switches within the grid-side converter. Namely, the subtransient currents can exceed the rated current by an order of magnitude. Therefore, it is necessary to use considerably larger power switches, and this could be the problem in large-power grid-side inverters. In addition to that, the hardware requirements also include a local means for the energy storage, such as the one in Fig. 8.10.

8.2.5 Negative Impedance

Many electrical loads have an active front-end converter. Electrical drive shown in Fig. 8.11 has the grid-side converter at the front-end, facing the mains. The motor-side inverter supplies the motor and draws the energy from the dc-link. The dc-link power depends on the motor speed Ω_m and torque T_{em}. Neglecting the losses, the dc-link power is equal to $\Omega_m \cdot T_{em}$. Thus, it is defined by the operating regime of the motor, and it remains essentially independent on the changes that originate from the grid. The dc-link power $\Omega_m \cdot T_{em}$ has to be supplied from the grid-side converter, shown in the left part of the figure. In cases where the motor brakes, the motor torque T_{em} changes the sign, and the power $\Omega_m \cdot T_{em}$ is negative. While braking, and assuming that the losses are negligible, the motor-side inverter regenerates into the dc-link, the grid-side converter draws the dc-power $-\Omega_m \cdot T_{em}$ from the dc-link, and it injects the same power into the grid. In the considered braking-operating mode, the grid-side converter acts as a source which supplies the energy to the ac grid.

The grid-side converter wave shapes the line currents. Without the lack of generality, it can be assumed that the electrical drive of Fig. 8.11 operates in motoring mode, with sinusoidal line currents, and with the current amplitude and phase which result in the power $\Omega_m \cdot T_{em}$ drawn from the grid and injected into the

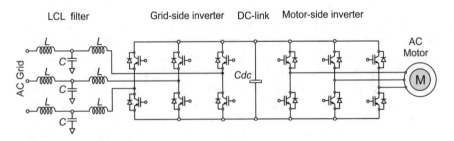

Fig. 8.11 Electrical drive with active front-end converter

dc-link. It is of interest to reinstate that the power $\Omega_m \cdot T_{em}$ does not depend on the line voltages and that the drive inner current control loops ensure that the grid-side converter operates with the motoring power $\Omega_m \cdot T_{em}$, notwithstanding the changes that originate from the grid.

Whenever the line voltages increase, the grid-side converter currents would decrease in amplitude in order to maintain the desired motoring power. The change of the current amplitude comes from the internal controller which maintains the dc-bus voltage. Namely, with an increase of the grid voltages, and assuming that the line-current amplitude remains initially unchanged, the power of the grid-side converter increases and exceeds the power consumed by the motor inverter. Therefore, the dc-bus capacitor gets charged, and the dc-bus voltage increases. In response to this increase, the internal controller which maintains the dc-bus voltage reduces the amplitude of the line-current references, and the grid-side power settles back to the steady-state value.

In a like manner, whenever the line voltages decrease, the line-current amplitude increases. From the above considerations, it is concluded that the electronic control of the converter structure of Fig. 8.11 contributes to negative equivalent dynamic resistance, as observed from the viewpoint of the grid. Namely, any positive change ΔU of the grid voltages causes a negative change ΔI of the line currents and vice versa. In other words, the equivalent dynamic resistance of the grid-side converter is negative, $R_{EQ} = \Delta U/\Delta I < 0$.

The negative dynamic resistance is also observed in non-regenerative, single-phase front-end converters such as the PFC rectifier shown in Fig. 8.12. The load is represented as the resistance connected across the dc-bus, but in most cases, the actual load is yet another switch mode power converter which draws the power from C_{DC} capacitor.

The line voltage $u_L(t)$ is used to calculate the reference for the line-current $i_L(t)$. The current amplitude is adjusted in order to draw the desired power from the grid. Corresponding waveforms are shown in the right part of the figure. The input power is maintained constant by means of the electronic control. Thus, when the line voltage increases, the amplitude of the line current decreases in order to keep the power constant. Any positive ΔU of the grid voltage causes a negative ΔI of the line currents, resulting in a negative equivalent dynamic resistance $R_{EQ} = \Delta U/\Delta I < 0$.

Negative dynamic resistance reduces the stability margin of the system. In cases where the grid has a large number of electronically controlled loads, their negative

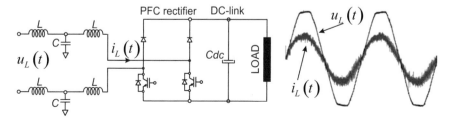

Fig. 8.12 Power factor corrector (PFC) converter with sinusoidal input current

dynamic resistance can bring the system to instability. In *R-L* networks, insertion of negative resistances reduces the equivalent time constant $\tau = L/R$. With significant impact of negative resistances, the constant τ assumes a negative value. In such cases, the initial disturbance does not decay as indicated by $\exp(-t/\tau)$, but increases instead, which drives the system to instability. Unfortunately, the problem of negative equivalent resistance in electronically controlled loads is not simple to solve. If the power is fully determined by the load conditions, it does not get affected by the line voltages. With electronic controls which maintain the power at the desired value, the currents are inversely proportional to the voltages, and the negative dynamic resistance is inevitable. In order to avoid the problem, it is necessary to add the means for the local energy accumulation, similar to the one shown in Fig. 8.10. With such buffer, it is possible to make the current proportional to the voltage transients over the short intervals of time, thus avoiding the negative resistance characteristics during transients. Even in this case, the steady-state values of the current are still inversely proportional to the voltages, but this circumstance does not affect the dynamic response.

8.3   DC-Bus Control and Droop Control

Static power converters used in conjunction with renewable power sources have generic topology that is illustrated in Fig. 8.13. It is of interest to discuss the problem of controlling the power conversion processes in the way that keeps the dc-bus voltage u_{DC} constant. At the same time, it is of interest to analyze parallel operation and the load sharing between several adjacent grid-side converters.

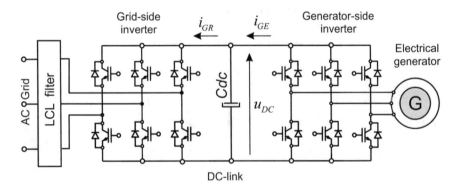

Fig. 8.13 Grid-side and generator-side converter

8.3.1 DC-Bus Control

In wind power sources, the power obtained from the generator is obtained in the form of variable-frequency ac currents. The generator-side inverter converts the power into the dc-bus power $u_{DC} \cdot i_{GE}$ (Fig. 8.13), which tends to charge the dc-bus capacitor C_{DC}. The grid-side inverter draws the dc-bus power $u_{DC} \cdot i_{GR}$, it performs the conversion of the dc voltage and current into line-frequency ac voltages and currents, and it injects the power into the grid. In order to keep the dc-bus voltage u_{DC} unchanged, it is necessary to keep $i_{GR} = i_{GE}$.

With solar plants, left part of the circuit in Fig. 8.13 remains the same, while the current i_{GE} comes from the dc/dc converters that process the dc power obtained from the strings of solar panels.

Control of the converter system in Fig. 8.13 has to maintain the dc-bus voltage constant in operation with a variable power input. There are two different ways to implement the control:

(i) Control of the generator (or the solar panel) is designed to search the operating point with the maximum power, and the current i_{GE} comes as the consequence of the maximum-power-point control. At the same time, the dc-bus voltage controller observes the dc-bus voltage, detects the dc-bus voltage error, and generates the current reference i_q for the grid-side inverter so as to drive the dc-bus voltage back to the reference value. The current i_q of the grid-side inverter determines the power injected into the ac grid. Thus, an increase in i_q will increase the current i_{GR}, which will, in turn, bring the dc-bus back into the balance $i_{GR} = i_{GE}$.

(ii) The power delivered to the grid from the grid-side inverter is calculated from the wind-generator-side variables (or, from the solar-panel-side variables). The grid-side power is controlled by setting the current i_q to the desired value. The reference value of i_q is calculated so as to bring the wind turbine (or solar panels) to the operating regime which delivers the maximum power. The current i_{GR} comes as the consequence of i_q. The changes in i_{GR} introduce variations of the dc-bus voltage u_{DC}. The dc-bus voltage controller sets the current references for the generator-side inverter (or, the solar-panel converter), and thus it alters the current i_{GE} in a way that brings the dc-bus back into the balance $i_{GR} = i_{GE}$.

The advantage of the first method is the possibility to change the generator-side converter currents and use them in the maximum-power-point search procedure (MPPT). With MPPT, the maximum-power operating point of the wind turbine (or, the solar panels) is found by the simple gradient search procedure which does not require additional measurements nor it require advanced a priori knowledge. The drawback of the first method are the fluctuations of the grid-side power. Namely, the i_q current of the grid-side converter and the grid-side power are the driving forces that come from the dc-bus-voltage controller, and they are used to maintain the dc-bus voltage u_{DC} at the desired value. Being the output of the closed-loop dc-bus-

voltage controller, the grid-side power and the current i_q exhibit significant variations that could not be acceptable in some grids.

The advantage of the second approach is a rather slow and smooth change of the
i_q current and the grid-side power. The drawback of this method is inability to
implement conventional MPPT control. Namely, the current i_{GE} and the power of the
generator-side inverter (or, the solar panels) cannot be stepped as required within a
conventional MPPT procedure. Instead, they are set by the output of the dc-bus-
voltage controller, and they have to secure the dc-bus equilibrium $i_{GE} = i_{GR}$. The
maximum point is obtained indirectly, by changing the power reference value for the
grid-side converter. In absence of the MPPT search procedure, this reference has to
be calculated on the bases of additional measurements and/or advanced a priori
knowledge on the wind turbine (or, the solar panels).

In cases where the system of Fig. 8.13 comprises the energy accumulation unit of
Fig. 8.10, there are more options to suite all the control and the performance goals, as
the equilibrium $i_{GE} = i_{GR}$ does not have to be maintained at all times.

8.3.2   Droop Control

In cases where several source-side converters operate in parallel, inject the ac power
into the common grid, and supply the same set of loads, it is often required that they
share the load proportionally to their rated power. When all the source-side converters have the same rated power, it is expected that each of them supplies an equal
share of the total load.

Equal power sharing between parallel-connected synchronous generators is
obtained by means of the power and frequency control means, often classified as the
primary, secondary, and tertiary control. The basic and the most immediate control
action of a synchronous generator that experiences a sudden load step is the speed
control action of the *governor*. While the electrical load increases the electromagnetic
torque, which decreases the rotor speed, the governor determines the amount of the
mechanical power supplied to the generator, namely, the moving torque obtained from
the hydroturbine, steam turbine, or another prime mover. The frequency of electrical
currents of the stator winding is related to the rotor speed. Thus, when an increase of
the load reduces the rotor, there is contemporary reduction of the frequency. The task
of the governor is to control the rotor speed, and this includes detecting the speed sags
and increasing the power/torque obtained from the prime mover.

The structure of the speed controller affects the static power-frequency characteristic of the generator, shown in Fig. 8.14. By using the conventional PI controller,
the presence of the integral action will contribute to the steady-state operation with
zero speed error. Namely, the steady-state rotor speed will remain equal to the
reference, and it will not change with the output power. In such case, the steady-
state characteristic in Fig. 8.14 will be a horizontal, constant-frequency line
(f = const.), inadequate for the grid-connected generator. With proportional speed
controller which does not include an integral action, the static characteristic will have

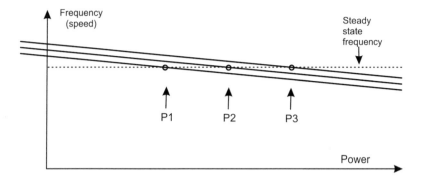

Fig. 8.14 Static characteristics of the governor-driven speed control

a finite slope $\Delta P/\Delta f$, determined by the proportional gain of the controller, as shown in Fig. 8.14. The steady-state characteristics $f(P)$ include a gradual reduction of the frequency/speed with an increase in the output power. Thus, the frequency *droops* with an increase of the output power.

The three curves plotted in Fig. 8.14 correspond to the three synchronous generators with proportional speed control and with a finite slope $\Delta P/\Delta f$. Due to imperfections in local settings and differences in the system parameters, the three curves cannot be identical, and they include small differences. The dashed horizontal line represents the constant-frequency operation that could be reached in the steady state, when all the generators operate with the same frequency. The steady-state powers P_1, P_2, and P_3 of the three generators are obtained at the intersection of the dashed, constant-frequency line and the corresponding $f(P)$ curve. Notice in Fig. 8.14 that the differences between P_1, P_2, and P_3 are determined by the slope $\Delta P/\Delta f$, and they increase with the (static) proportional gain of the speed controller. In cases with very large proportional gains, the differences can be very large as well. In an extreme case, which corresponds to the use of speed controllers with the steady-state speed error equal to zero, some of the parallel-connected generators will run with no load, while others will be overloaded. In order to obtain an equal power sharing of parallel-connected synchronous generators, the governor speed controllers include a significant frequency drop in cases where the output power increases.

The droop control, illustrated in Fig. 8.14, implies a significant drop in the A typical slope of the $f(P)$ curves in Fig. 8.14 is close to 4%. Namely, the power step from zero to the rated power (100%) produces the frequency change of $0.04 \cdot 50 = 2$ Hz. That is, in cases without any other corrective measures, and with the governor speed controller running at 50 Hz and zero-power, the rated load power will produce the frequency drop from 50 Hz down to 48 Hz. Practical load steps and sudden generator outages produce considerably smaller frequency changes. The most usual frequency transients do not exceed 0.1 Hz. In order to bring the line frequency back to the reference (50 Hz), there are secondary and tertiary frequency/power control actions, which are not discussed in this book. The governor speed controllers operate in the timeframe from several tens of milliseconds to several seconds. The speed control action is based on a certain power margin available from the prime mover and the generator itself. If the prime mover and the generator are running close to their maximum power rating, there is next to nothing that the speed controller can do in order to maintain the speed.

In wind power plants and solar power plants, an attempt to use the power obtained from the prime mover as a driving force of the frequency controller will result in the wind turbine or solar-panel operation off the maximum power point, thus reducing the harvested energy. The converter pairs, such as the one in Fig. 8.13, have the task of maximizing the energy harvesting from the wind turbines and solar panels and injecting the obtained power into the ac grid. Thus, the power injected into the grid by any of these grid-side converters is not the matter of choice. Neglecting the losses, it has to be equal to the maximum power that is available from the corresponding renewable source at any given time. In cases where several renewable sources with their grid-side converters operate in parallel and close to each other, the output power supplied by each of them remains equal to the power obtained from the corresponding wind turbine or solar panel. Thus, there is no need for any of these parallel-connected grid-side converters to share the total power. Considering a wind farm with ten parallel-connected, equally rated converter sets of Fig. 8.13, it is not required nor it happens that each of them supplies 1/10 of the total power. Instead, each converter set will supply the power obtained from its own wind turbine, and those powers are rarely the same.

The operation of source-side power converters is different in ac microgrids. In an attempt to provide reliable source of electrical energy to critical loads, to supply the loads in areas where the main grid is not available, and to increase the reliability of the power supply, microgrids and distributed generation are becoming more popular. At the same time, national emergency operations and defense systems often require a scalable ac grid with sufficient robustness to ride through some radical changes. The equivalent inertia of all the generators connected to a typical microgrid is usually low, in particular when the grid has to be portable. For this reason, the kinetic energy kept in rotating masses is also low. Large power transients result in large frequency changes, and such changes could provoke a loss of synchronism and instability. Stability is jeopardized by an attempt to maintain the line-frequency constant even in cases where the load changes are large. In such cases, it is of uttermost importance to adjust the droop-control mechanism so that they could accommodate sudden and large transients. A droop control, illustrated in Fig. 8.14, related the power transients to the frequency change. The speed/frequency controllers with proportional gain improve the system stability reducing the impact of sudden power imbalances, providing in this way the necessary escape time for the system to prevent the loss of synchronism.

A considerable number of sources in microgrids are the electronically controlled sources with grid-side power converters. Many of them draw the power from batteries, fuel cells, micro-turbines, and gas turbines, the prime energy sources that can provide the power output that can be changed at will, thus providing the grounds for the implementation of the droop control. Based upon the measurements of the local power and frequency, it is possible to adjust the operating point of each grid-side converter according to the droop-control power-frequency rule, illustrated in Fig. 8.14. By implementing the droop control, the frequency control in a microgrid is not stiff, and a finite $\Delta P/\Delta f$ slope contributes to the system stability. A more thorough analysis could prove that the frequency drop of the curves in Fig. 8.14 has to be larger in cases where the rotating masses are particularly low.

8.4 DC-Bias Detection and Suppression

Grid-side inverters comprise electronically controlled switching bridges which convert the dc power, obtained from the dc-link into ac power, injected into the grid by means of digitally controlled and wave-shaped ac currents. They also operate in other directions, converting the ac power into dc power. The voltages are generated by the pulse width modulation, and the current control relies on dedicated current sensors. There is no intention to inject a dc current into the ac grids. Yet, due to imperfection of sensors, actuators, and controllers, it is possible for the grid-side inverter to introduce not only the ac currents into the grid but also a small, parasitic amount of dc current, usually called the dc-bias.

Injection of dc currents jeopardizes the operation of line-frequency power transformers. In addition to that, the presence of dc-bias can produce adverse and harmful consequences in many electrical loads. With an increased number of grid-side power converters, negative consequences of the dc-bias are emphasized evermore. Due to a very low equivalent resistance of ac grids, even some very low values of the dc-bias voltages can produce serious harmful effects in power transformers and put them out of the service. For this reason, the dc-bias has to be measured with a very high precision, providing at the same time the means for active compensation and absorption of the bias. Namely, with the proper measurement of the bias, the neighboring grid-side converters can provide the counter-injection that would compensate the bias and bring the parasitic dc voltage to zero.

The grid codes for the grid-side power converters introduce the limit for the dc injection to 1/200 (0.5%) of the device rated current. Currently, the relative share of electronically controlled converters is not high. In a hypothetical case where all the sources and loads are connected through electronically controlled interfaces, it could be expected that the parasitic dc current in the secondary windings of distribution transformers reaches 0.5% of each transformer rated current. Typical magnetizing current of distribution transformers is well below 1%. Therefore, just 0.5% of the dc-bias would drive the transformer into deep saturation, triggering the protections and putting the transformer out of the service.

It is of interest to get an insight into acceptable dc-bias voltages in 0.4 kV ac grids. The dc-bias component of the magnetizing current of distribution transformer depends on the bias voltage and on the resistance of the transformer windings. This resistance is usually below 0.5% of the ratio U_{nom}/I_{nom}. Assuming that both the relative value of the winding resistance and the relative value of the magnetizing current are 0.5% and that the dc-bias voltage on 0.4 kV side of the transformer is

1 mV, it is possible to calculate the effects of the bias. With 1 mV being just 0.001/ 400 = 0.00025% of the ac voltages, the dc-bias current in the magnetizing branch will reach 0.001/400/0.005/0.005 = 10% of the rated magnetizing current. As a consequence, the offset is introduced into *B-H* magnetizing curve of the core, and it caused half-cycle saturation. This phenomenon affects the half cycles of the magnetizing current that are aided by the bias. These half cycles have considerably larger peaks of the magnetizing current, with a consequential increase in the reactive power, stray losses, and iron losses in the transformer.

The problem of dc-bias injection gets worse with perpetual increase of the share of grid-side power converters and with the use of new, high-efficiency distribution power transformers. Novel power transformers make use of advanced ferromagnetic materials, and they have even lower values of the winding resistance and the magnetizing current. Therefore, they are even more susceptible to the dc-bias. The need to suppress the dc-bias far below 1 mV in 0.4 kV grids calls for the sensing methods with elevated precision. It is necessary to detect very small dc offsets superimposed on 10^5–10^6 times larger ac voltages, and this task cannot be achieved with currently available sensing devices. The most promising methods for reading the dc-bias rely in a small, saturable sensing reactor connected in parallel with the biased ac voltage. When an ac-exited reactor obtains a dc-bias, the magnetizing current of the reactor gets distorted. Distortion of the reactor current I_m is be used to detect the polarity and magnitude of the dc-bias. A rather simple and straightforward approach includes the comparison of the positive and negative peak of the magnetizing current. With dc-bias, one of these peaks is going to prevail. The difference between the peak values provides the sign and magnitude of the dc-bias, and this signal can be used to detect and suppress the dc-bias in ac grids. An alternative approach in calculating the dc-bias from distorted magnetizing current is to consider the spectral content. The presence of the dc-bias in a saturable core gives rise to low-order even harmonics. Their amplitude and phase provide an accurate information on the dc-bias.

In this section, the analysis of the dc-bias sensing proves that the precision of the sensor largely depends on the form of the sensing reactor. Further on, it is shown that the even harmonics method provides considerably better results than the peak-detection method. The section ends with the sample design of the dc-bias suppression controller. The considerations outlined in the section are supported by experimental results, obtained with the sample grid-side power converter.

8.4.1 Sensitivity of Distribution Transformers to DC-Bias

The dc-bias introduced by grid-side power converters introduces the problems that are mostly observed in distribution power transformers [1]. When the load-side and source-side converters are connected to 0.4 kV grid, the dc-bias offsets the magnetizing current at the secondary winding of distribution transformers. Sensitivity of the transformer to the parasitic dc-bias applied across the secondary winding has to be studied for oil-immersed transformers [2], cast-resin dry-type transformers [3], and amorphous alloy core transformers [4]. The problems caused by the dc-bias increase in cases where the dc-bias current reaches a significant fraction of the rated magnetizing current. Therefore, an indicator of the problem severity can be expressed in terms of the ratio of the bias current I_{mDC} and the rated magnetizing current I_{mnom}. When the ratio I_{mDC}/I_{mnom} reaches 0.05 (5%), the core losses can increase by 22% [5]. The parasitic dc-bias current I_{mDC} is determined by the ratio between the parasitic dc voltage across the secondary winding (U_{DC}) and the resistance of the secondary winding (R_2). With $I_{mDC}/I_{mnom} = U_{DC}/(R_2 I_{mnom})$, the problems caused by the dc-bias increase when the winding resistance is low, and they also increase when the transformer magnetizing current is low. The sensitivity to the dc-bias thus depends on the winding resistance and the rated magnetizing current of the transformer.

Relative values of the winding resistances and relative values of the rated magnetizing currents are given in Fig. 8.15 for conventional oil-immersed distribution transformers up to $S_n = 2500$ kVA. The base value for the winding resistance is the ratio $Z_{nom} = U_{nom}/I_{nom}$ of the rated voltages and currents. The base value for the excitation current is the rated current of the transformer. Both relative values decrease with an increase in the rated power S_n of the transformer. For 2500 kVA transformers, the rated magnetizing current is lower than 1% ($I_{mnom} < 0.01$ p.u.), while the relative winding resistance drops below 0.5% ($R_2 < 0.005$ p.u.). In order to obtain the information on the dc-bias sensitivity, the figure includes the ratio I_{mDC}/I_{mnom} which is obtained assuming that the dc-bias voltage of $U_{DC} = 1$ mV is brought

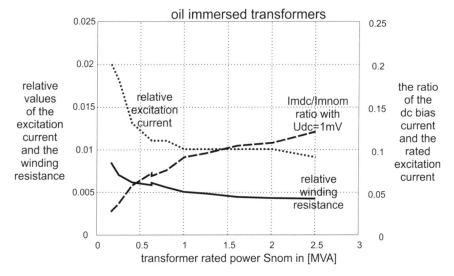

Fig. 8.15 Relative values of the winding resistances and magnetizing currents for oil-immersed distribution transformers up to 2500 kVA. The ratio between the parasitic dc current I_{mdc} and the rated magnetizing current I_{mnom} is given for the assumed parasitic dc-bias of 1 mV across the secondary winding

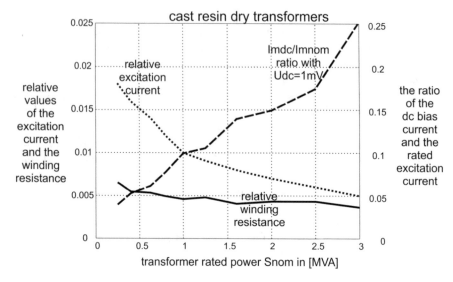

Fig. 8.16 Relative values of the winding resistances and magnetizing currents for dry distribution transformers up to 3000 kVA. The ratio between the parasitic dc current I_{mdc} and the rated magnetizing current I_{mnom} is given for the assumed parasitic dc-bias of 1 mV across the secondary winding

across the secondary winding. The ratio I_{mDC}/I_{mnom} increases with the rated power of the transformer (i.e., it increases with the transformer size). Thus, with an oil-immersed transformer of 2500 kVA, and with the dc-bias voltage of 1 mV, the parasitic dc current I_{mDC} reaches 12% of the rated magnetizing current. According to reports in [5], this amount of dc-bias results in a considerable increase in the core losses and the reactive power.

In Fig. 8.16, similar results are plotted for the dry-type cast-resin distribution transformers up to 3000 kVA. The winding resistances of dry-type transformers are lower than the resistances plotted in Fig. 8.15. The same holds for relative values of the rated magnetizing current. For the dry-type cast-resin distribution transformer of 3000 kVA, the parasitic dc-bias voltage of 1 mV introduces the parasitic dc-bias current equal to 25% of the rated magnetizing current, thus doubling the value obtained with oil-immersed transformers.

Distribution transformers with their cores made of amorphous ferromagnetic materials [4] have lower iron losses and considerably lower excitation power. With the amorphous alloy core transformer, the ratio I_{mDC}/I_{mnom} is 5 times higher than with the conventional core dry-type transformer of the same rated power. The transformer sensitivity to the dc-bias is increased by the same amount.

The results shown in Figs. 8.15 and 8.16 can be used to obtain a rough estimate of the parasitic dc-bias voltage that would increase the core losses by 20%. These calculations rely on the results published in [5]. In oil-immersed transformers rated 2500 kVA, it takes 420 μV of parasitic dc-bias voltage to obtain an increase in the core losses of 20%. For dry-type cast-resin transformer of 3000 kVA, corresponding amount of parasitic dc-bias is only 200 μV. The bias limit is even lower for the transformers with amorphous alloy cores, where the measurement and suppression of parasitic dc-bias has to reach precision of 100 μV.

8.4.2 Peak-Detection Methods

In most cases, detection of the parasitic dc-bias relies on parallel-connected saturable reactor. The magnetizing current of the reactor gets affected by the dc-bias. The bias introduces the difference between positive and negative peaks of the reactor current, and it introduces even harmonics into the current waveform. The parallel sensing reactor can be used in sensing circuits, where the main goal is to measure the amount of dc-bias (Fig. 8.17). At the same time, the sensing reactors can be used as the feedback device in dc-bias suppression systems (Fig. 8.18).

In Fig. 8.17, the sensing reactor has the main winding and the auxiliary, compensation winding. The main winding is connected across the ac grid terminals. For the proper operation of the peak-detection method, the ac voltage has to be filtered by means of a low-pass LC filter which removes the noise and spurious spikes that

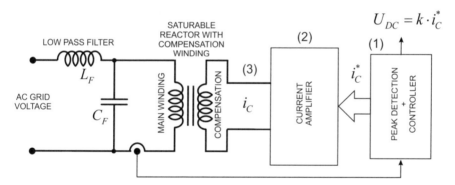

Fig. 8.17 Parallel reactor with compensation winding, peak-detection circuit, closed-loop zero-bias controller, and an amplifier

Fig. 8.18 Parallel reactor with grid-side converter which injects the dc offset U_{GSC} to compensate the parasitic dc-bias U_{DCG} of the grid and to obtain the zero-bias operation of the reactor with $U_{R0} = 0$

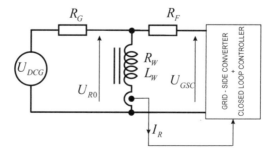

could impair detection of the peak values in the reactor current. The parasitic dc current in the main winding is equal to $I_{mDC} = U_{DC}/\Sigma R$, where ΣR is the sum of all the series resistances. The offset I_{mDC} affects the magnetizing current I_m in the main winding, and the distortion of the current provides the means for the dc-bias detection. In block (1) of the figure, the peak-detection circuit calculates the difference between positive and negative peaks. On the bases of this feedback, the closed-loop controller outputs the current reference $i_C^{\,*}$ for the compensation winding. The current amplifier (2) injects the desired current i_C into the compensation winding (3). The current reference $i_C^{\,*}$ acts against the dc-bias I_{mDC}. In steady state, the closed-loop controller brings the reactor core in zero-bias conditions, where the ampere-turns of the current I_{mDC} correspond to the ampere-turns of the current i_C. Thus, the signal $i_C^{\,*}$ gets proportional to the dc-bias voltage U_{DC}, and it can be used to output the signal proportional to U_{DC}. The circuit in Fig. 8.17 is used as the sensor which provides the information on the dc-bias, but it does not provide any corrective measure that would suppress the bias in the grid.

In Fig. 8.18, the reactor current signal is brought into the grid-side power converter that is connected to the grid at the same connection point as the sensing reactor. The controller that runs within the grid-side power converter processes the reactor current and obtains the dc-bias feedback, and it calculates the desired offset voltage U_{GSC} that has to be generated at the output terminals of the grid-side converter in order to compensate the grid bias U_{DCG} and bring the dc-bias U_{R0} down to zero. In steady state, any grid-side power converter equipped with the circuit of Fig. 8.18 would ensure that the dc-bias across the grid-connection point gets equal to zero. Thus, instead of injecting the parasitic dc offset, the grid-side power converters would absorb and compensate the bias that originates from the grid.

Peak-detection dc-bias sensors [6] use the difference ΔI between positive and negative peaks of the reactor current I_m. The peaks are obtained by sampling I_m at zero crossing instants of the applied voltage (Fig. 8.19). The difference *IMAX-IMIN* is proportional to the dc-bias within the reactor core. The relevant samples are acquired in each half period of the line voltage. The use of only one sample in each half period makes the sensing method prone to the noise and offset in the signal processing chain. In order to suppress the noise that originates from the grid, it is necessary to introduce an *LC* low-pass filter between the noise source and the

Fig. 8.19 Magnetizing current of the toroidal sensing core exposed to the dc-bias and to the ac excitation

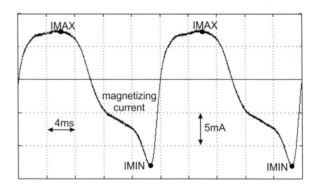

Table 8.1 Relative peak difference $\Delta I/I_{peak}$ measured with I_{mDC} that ranges from 0 up to 0.125 I_{mnom}. The core has $I_{mnom} = 8$ mA and S_{FE} of 3 cm^2. I_{peak} corresponds to the peak value of $I_m(t)$

U_{AC} [V]	150	170	200	220	264
$I_{mDC} = 0.125\, I_{mnom}$	0.18	0.31	0.5	0.87	0.16
$I_{mDC} = 0.04\, I_{mnom}$	0.0096	0.067	0.1245	0.1988	0.0413
$I_{mDC} = 0.02\, I_{mnom}$	−0.0004	0.033	0.0572	0.0991	0.0194
$I_{mDC} = 0$	−0.003	0.008	0.0105	0.0209	0.0114

reactor. An $L_F C_F$ filter is included in Fig. 8.17. The sensor reported in [6] has the reactor with $R_W = 29$ Ω and a choke L_F with $R_{LF} = 56$ Ω. The equivalent resistance is $\Sigma R = R_W + R_{LF} = 85$ Ω. With larger resistance, one and the same dc-bias U_{DC} produces lower dc currents I_{mDC}, reducing in this way the sensitivity. Therefore, it is desirable to use detection methods that are not sensitive to the offset and noise, since they do not require any LC filtering, and their equivalent series resistance is considerably lower.

The smallest detectable current I_{mDC} which produces reliable information is denoted by $I_{mDC(min)}$. Various noise and offset voltages result in the measurement error ΔI_{mDC}. The ratio $I_{mDC(min)}/\Delta I_{mDC}$ defines the signal-to-noise ratio. With peak-detection methods, the parasitic dc current I_{mDC} affects the flux in the core and introduces distortion in the magnetizing current I_m, creating the difference ΔI between the peaks. The difference ΔI depends on the ratio $Q = I_{mDC}/I_{mnom}$, that is, it is easier to detect the offset I_{mDC} in cases where the rated magnetizing current I_{mnom} is lower. The impact of Q on detectable ΔI is experimentally studied on a sample reactor which has the rated magnetizing current I_{mnom} and the cross section S_{Fe} given in [7]. Assuming that the average value of the peak current is I_{peak}, the relative values of the peak difference can be expressed as $\Delta I/I_{peak}$. Corresponding values are given in Table 8.1 for a range of ac voltages and I_{mDC} currents. It is of interest to estimate the minimum detectable $I_{mDC(min)}$ and the minimum operable ratio Q_{min}. For the offset current I_{mDC} equal to zero, and for $U_{AC} = 220$ V, the value $\Delta I/I_{peak}$ is equal to 0.0209. By linearizing the function $\Delta I/I_{peak} = f(I_{mDC})$ at the given operating point, one obtains the value of I_{mDC} required to bring ΔI to zero, and this value is $I_{mDC0} = -I_{mnom}/200$. The value I_{mDC0} represents the measurement error ΔI_{mDC}. By taking $I_{mnom}/200$ as the minimum detectable current $I_{mDC(min)}$, detection threshold is $Q_{min} = 1/200$. With the resistances given in [7], the voltage detection threshold $U_{DC(min)}$ is 2.2 mV, which is close to $\Delta U_{DC} = 3$ mV, reported in [7] for $U_{AC} = 220$ V.

In operation with variable ac voltages, the peak-detection methods exhibit a reduced precision due to their need to operate next to the knee of the B-H curve. In [7], the errors ΔU_{DC} are plotted for ac voltages that change from 170 V up to 260 V. The minimum values of the error are obtained for ac voltages in the middle of the range. The error ΔU_{DC} increases toward 170 V, as well as for the voltages that approach the upper end of the range. The error reaches 15 mV for $U_{AC} = 260$ V. For the given offset current I_{mDC}, the values of $\Delta I/I_{peak}$ in Table 8.1 reduce as U_{AC}

moves away from 220 V in either direction. Saturation at 264 V may alter the waveform $I_m(t)$ and mask the signal ΔI. At 150 V, the operating point in B-H plane returns to the linear region, where the difference ΔI is less pronounced. In the subsequent sections, it will be shown that the even harmonics-based detection method maintains a high precision over a wide range of ac voltages.

8.4.3   Optimum Form of the Core

The smallest detectable bias voltage $U_{DC(min)}$ depends on the total resistance ΣR of the primary circuit and the minimum detectable bias current $I_{mDC(min)}$:

$$U_{DC}^{min} = (\Sigma R) \cdot I_{mDC}^{min} = (\Sigma R) \cdot I_{mnom} \cdot Q_{min} \qquad (8.34)$$

With peak-detection methods, the minimum ratio $Q_{min} = I_{mDC(min)}/I_{mnom}$ is close to 1/200. Thus, precision of the dc-bias sensor depends on the product $(\Sigma R)I_{mnom}$. In order to increase precision, it is necessary to minimize the winding resistance and to minimize the rated magnetizing current. Given the need to minimize the product $\Sigma R \cdot I_{mnom}$, the best results are obtained with toroidal-shaped cores. When designing a low-cost dc-bias sensor, it is of interest to consider a standard, commercially available core that uses most common iron laminations. Design consists in selecting commercially available core, selecting the wire, and determining the number of turns. Dimensions, weight, and electrical parameters of several commercially available toroidal transformers with $20VA < S_n < 500VA$ are listed in Table 8.2. The primary winding of these transformers is made for 220 V, 50 Hz. The rated magnetizing current I_{mnom} and the primary resistance are given for all the samples. The product of the primary resistance R_{prim} and the current I_{mnom} is given in the rightmost column, labeled as $R3I$.

The sensitivity is inversely proportional to $R3I$ product (8.34). Therefore, from Table 8.2, it is beneficial to use larger cores. The core sensitivity can be roughly doubled by selecting the core with 10 times larger rated power. It takes roughly 10 time larger weight of the core to increase the sensitivity some 3 times. Given the

Table 8.2 Properties of toroidal, single-phase, line-frequency transformers with POWER CORE H 105-30 iron sheets by thyssenkrupp AG. R_{PRIM} stands for the primary resistance (220 V side)

S_n [VA]	S_{FE} [mm^2]	D_{ext} [mm]	m [kg]	I_{mnom} [mA]	R_{prim} [Ω]	$R3I$
30	250	60	0.43	1.42	70	99
50	300	70	0.63	2	31	62
80	300	80	0.85	3.5	24	84
150	400	100	1.5	4.5	11	49
200	600	100	1.75	6.7	8	53
300	800	100	2.5	8.2	4.1	33
400	750	120	4	10	3.2	32
500	1000	120	4.3	12	2.3	27

same cross section S_{FE}, better results are obtained with the core having a lower diameter. Thus, some core forms have the potential of reaching better precision than the others.

The main drawback of using an off-the-shelf toroidal transformer is a larger resistance of the main winding. The resistance is larger due to the fact that roughly 50% of the available winding space is used for the secondary winding. It is also possible to use an off-the-shelf device and rewind it, using most of the winding space for the main winding, thus halving its resistance. The winding space dedicated to the auxiliary, compensation winding can be rather small, due to a low number of turns that carry a relatively small compensation current i_C (Fig. 8.17).

Better results are obtained by providing the toroidal core of an optimum form and making a dedicated winding. In further discussion, the impact of the core form on the sensing precision is studied in order to find the optimum core form, which provides the best precision for the given core size.

For the core with the winding resistance R_W, for the rated magnetizing current I_{Rnom}, and for the given ratio $Q = I_{mDC(min)}/I_{Rnom}$, the value of U_{DC}^{min} is obtained as $(R_W I_{Rnom})Q$, as shown in (8.34). It has been shown (Table 8.2) that the precision increases with the core size. Still, the overall size of the sensing reactors is limited due to practical reasons. Therefore, it is necessary to find the optimum form of the reactor core which provides the best precision for the given size of the core.

For toroidal core with external diameter D_e, internal diameter D_i, and height W, the core form can be expressed in terms of the form factors $d = D_i/D_e$ and $w = W/D_e$. The core volume and weight are

$$V = \left(D_e^2 - D_i^2\right) \cdot W \cdot \frac{\pi}{4}, m = V \cdot 7650 \frac{\text{kg}}{\text{m}^3}. \tag{8.35}$$

The volume V can be expressed in terms of the external diameter D_e and the form factors d and w. The core with one and the same volume (and weight) can be obtained with different factors d and w. The choice of the form factors affects the parameters R_W and I_{Rnom} and the product $R_W I_{Rnom}$. In this way, the form of the core has a direct impact on the overall sensing precision. For this reason, it is of interest. Given the core volume, it is of interest to derive the form factors that minimize the product $R_W I_{Rnom}$. The current I_{Rnom} depends on the peak flux density B_{max} and peak strength H_{max} of the magnetic field, on the number of turns N_T, and on the core geometry:

$$I_{Rnom} = H_{max} \cdot \frac{\pi(D_e + D_i)}{2} \frac{1}{N_T} \frac{\sqrt{2}}{2} \tag{8.36}$$

The winding resistance R_W depends on N_T, on the wire cross section S_{Cu1}, and on the average length of one turn L_{av} (8.37), while the number of turns depends on the core cross section S_{Fe}, on the ac voltage U_{AC}, and also on the line-frequency ω_{AC} (8.38).

$$R_W = \frac{N_T L_{av}}{S_{Cu1}} \frac{1}{\sigma_{Cu}}, \sigma_{Cu} = 57 \cdot 10^6 \frac{S}{m}. \tag{8.37}$$

$$N_T = \frac{U_{AC}\sqrt{2}}{\omega_{AC}} \frac{1}{S_{Fe}B_{max}} = \frac{U_{AC}\sqrt{2}}{\omega_{AC}B_{max}} \frac{2}{W \cdot (D_e - D_i)} \tag{8.38}$$

The feasible cross section S_{Cu1} of the copper wire is obtained by multiplying the available winding area W_A by the copper fill factor α_W and dividing the result by the number of turns N_T. The shuttle of the winding machine has to pass through the round hole, namely, the internal diameter D_{iF} of the finished reactor, including the windings. The minimum value of D_{iF} is $\beta \cdot D_i$, where β is defined by specific properties of the winder equipment, and it is usually close to 1/2. With $D_{iF} > \beta \cdot D_i$, the largest winding thickness is $W_T = (1-\beta) \cdot D_i/2$, while the available winding area through the central opening of the toroidal core is $W_A = (1-\beta^2) \cdot \pi \cdot D_i^2/4$. Considering the copper fill factor of α_W, the wire cross section is (8.39), and the average length L_{av} of one turn is (8.40), and the winding resistance is in (8.41).

$$S_{Cu1} = \alpha_W (1 - \beta^2) \pi D_i^2 / (4N_T) \tag{8.39}$$

$$L_{av} = D_e + (1 - 2\beta)D_i + 2W \tag{8.40}$$

$$R_W = \frac{4N_T^2[D_e + (1 - 2\beta)D_i + 2W]}{\alpha_W(1 - \beta^2)\pi D_i^2} \frac{1}{\sigma_{Cu}} \tag{8.41}$$

Finally, the product $R_W I_{Rnom}$ can be expressed as a function of the form factors d and w:

$$R_W I_{Rnom} = f_x(d, w) = \frac{k}{D_e^2} \frac{(1 + d)[1 + (1 - 2\beta)d + 2w]}{d^2 w(1 - d)}, \tag{8.42}$$

where

$$k = (4U_{AC}H_{max}) / [\alpha_W \sigma_{Cu} (1 - \beta^2)\omega_{AC}B_{max}]. \tag{8.43}$$

The expression (8.42) defines the $R_W I_{Rnom}$ product as the function $f_x(d,w)$. The maximum sensing precision for the given core volume is obtained when the function $f_x(d,w)$ reaches the minimum value. For this to achieve, it is necessary to find the form factors which provide the smallest value of $f_x(d,w)$, that is, the largest value of $1/f_x(d,w)$. This value can be obtained by a parametric search in d-w plane. The results are plotted in Fig. 8.20, with $1/(R_W I_{Rnom})$ on vertical axis. The form factors d and w are given on x and y axis. The maximum sensitivity is obtained for $d_{opt} = 0.76$ and $w_{opt} = 0.324$. The results of Fig. 8.20 show that the proper selection of the form factors can increase precision by nearly two times. The increase in precision is confirmed by experimental results, obtained with the optimum-form core of Fig. 8.21. The results are summarized in Table 8.3. Although the conventional-form core of [8] has a larger mass and size, the optimum-form core offers considerably better precision (65 μV versus 100 μV).

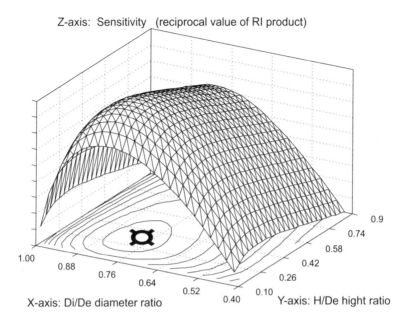

Fig. 8.20 The core sensitivity $1/(R_W I_{Rnom})$ is calculated from (8.42) as a function of the form factors, assuming that the core weight remains constant

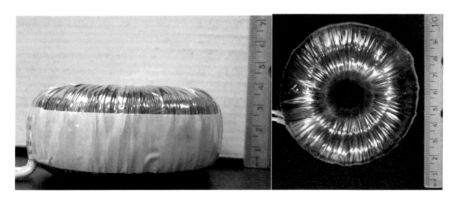

Fig. 8.21 Sample sensing reactor made on an optimum-form iron core

Table 8.3 Comparison between the sensing reactor of [8] and an optimum-form reactor: The core weight (m_{FE}), the copper weight (m_{Cu}), the total weight (m_{tot}), the sensing error (ΔU_{DC}), magnetic induction (B_m), and diameter ratio

Sensing reactor	m_{Fe} [kg]	m_{Cu} [kg]	m_{tot} [kg]	ΔU_{DC} [µV]	B_m [T]	D_e/D_i
Reactor of [8]	0.408	0.506	0.914	< 100	1.5	1.75
Sample core of Fig. 8.21	0.305	0.420	0.725	< 65	1.505	1.48

8.4.4 Detection Based on Even Harmonics

The current of the sensing reactor gets affected by the dc-bias. Distortion of the reactor current provides an indication on the sign and magnitude of the dc-bias. The current gets processed by the sensing algorithm to extract the dc-bias feedback information. The peak-detection methods rely on time-domain samples of the current. When operating with the rated ac voltage, the peak-detection methods offer precision better than 3 mV [6, 7]. The method reported in [8] is based on detection of even harmonics of the reactor current. The amplitude of even harmonics increases with the dc-bias, while the phase of even harmonics indicates the sign of the bias. The even harmonics-based method of [8] reaches precision better than 100 μV. Therefore, it is of interest to describe the signal processing chain and the implementation of this method.

The even harmonics of the reactor current are created in an interaction of the dc-bias and the nonlinearity of the ferromagnetic material. At the same time, certain amount of even harmonics can originate from the grid. These harmonics can produce the even harmonics of the reactor current that are not related to the bias. In order to distinguish between the two, it is necessary to detect the phase of the even harmonics. Namely, the even harmonics that originate from the interaction between the dc-bias and the nonlinearity of ferromagnetic material have the specific phase with respect to the first harmonic. For several low-order even harmonics that originate from the bias, their phase [8] remains rather constant for a range of operating conditions. Thus, contemporary presence of these harmonics provides a unique and conclusive indication on the presence of the dc-bias.

It is of interest to compare the relevant phase shifts for two different sensing cores. The even harmonics obtained with the reactor of [8] and harmonics obtained with the optimum-form reactor (Fig. 8.21) are given in Table 8.4. The values φ_2 .. φ_8 stand for the phase displacement of the given harmonic with respect to the first harmonic. The last row in the table provides the difference between the phase shifts obtained with the two cores. For both cores, the phase-shift φ_2 remains close to zero, within a rather narrow strip. The phase φ_4 changes between the two cores by more than 2 radians. If the signal processing chain tuned for one of the cores gets used for the other core, the error of 2 radians impairs the operation of the zero-bias tracking loop. Large phase differences are also found with the sixth and eighth harmonic in the table.

The results of Table 8.4 prove that the phase shifts of the fourth, sixth, and eighth harmonic depend on the specific core shape and core material. For this reason, the

Table 8.4 Phase shifts of the four even harmonics obtained with the sensing reactor of [8] and with the optimum-form reactor

Core	φ_2	φ_4	φ_6	φ_8
(1) sensing reactor of [8]	−0.08	0.48	1.05	1.54
(2) optimum-form reactor	0.06	3.12	0.06	3.01
(2)-(1)	0.14	2.64	−0.99	1.47

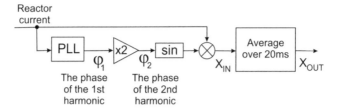

Fig. 8.22 Detection circuit which relies on the second harmonic – demodulator

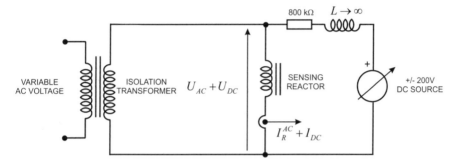

Fig. 8.23 The circuit for testing the sensing reactor in conditions with variable ac voltage and with variable dc-bias

use of the considered harmonics for the feedback purposes requires an off-line measurement of their phase shifts and subsequent calibration of the relevant parameters. For simplicity, it is of interest to use the sensing algorithm that relies on the second harmonic only, avoiding in this way any off-line tuning and calibration.

Demodulation circuit is given in Fig. 8.22. Reactor current is brought to the PLL circuit which extracts the phase φ_1 of the fundamental component of the current (the first harmonic), and it calculates the expected phase of the second harmonic φ_2, which corresponds to the second harmonic component produced by the dc-bias. This phase is used by the demodulator to obtain the signal x_{IN}. Raw demodulated signal x_{IN} is averaged in each period of the line voltage to obtain the output signal x_{OUT}. Therefore, the time interval of $T_{SPL} = 20$ ms is the sampling time of the closed-loop dc-bias controller.

The sample reactor is tested within the circuit of Fig. 8.23. The variable ac voltage is obtained from an autotransformer. The isolating transformer is used to isolate and protect the measurements of any additional parasitic dc-bias that could come of the ac grid. The dc-bias required for the test is obtained from a variable dc source (U_{DCIN}). In order to decouple the dc source from the ac voltages and currents, it is connected across a very large series impedance. More details about the testing setup, procedures, and results are provided in [9].

The reactor is tested for ac voltages between 160 V and 250 V and with dc-bias currents that range from −40 μA up to +40 μA. Corresponding dc voltages across

Table 8.5 Long-term average values of x_{OUT}, obtained with $I_{DC} = 0$, and the calculated values of the sensing error ΔU_{DC}

U_{AC} [V]	160	180	200	220	240	250
$|x_{OUT(av)}|$ [μA]	3.85	2.78	1.71	2.7	3.08	3.68
$\Delta I_{DC} = |x_{OUT}|/S_2$ [μA]	1.8	1.3	0.8	1.26	1.44	1.72
$\Delta U_{DC} = R_W \, \Delta I_{DC}$ [μV]	39.7	28.7	17.6	27.8	31.8	38

the sensing reactor are determined by the winding resistance, and they change from -884 μV up to $+884$ μV. The signal acquisition, oversampling, and filtering are performed by the DMA-driven ADC peripheral of the DSP controller. Experimental results with demodulated second harmonic $x_{OUT}(I_{DC})$ change in a linear manner [9] with an average slope of

$$S_2 = \frac{\Delta x_{OUT}}{\Delta I_{DC}} = 2.14 \frac{\mu A}{\mu A} \tag{8.44}$$

Several curves $x_{OUT}(I_{DC})$ were obtain by repeating the sweep of the dc-bias current between the limit values. All these quasi-linear curves $x_{OUT}(I_{DC})$ have their zero crossing (the value of I_{DC} that provide $x_{OUT} = 0$) close to $I_{DC} = 0$, in a ± 3.5 μA wide strip, as shown in Fig. 9 of [9]. When multiplied by the resistance R_W of the sensing reactor winding, corresponding voltage error is $\Delta U = 78$ μV. If the error is calculated for each of the curves, and the average value ΔU_{av} of the error is obtained for a series of consecutive measurements, the value ΔU_{av} gets even smaller.

In order to estimate the long-term steady-state errors, a number of consecutive measurements of x_{OUT} is taken in [9], all of them obtained while keeping the reactor in zero-bias conditions ($I_{DC} = 0$). The sensing error expressed in terms of the dc-bias current (ΔI_{DC}) is obtained as x_{OUTav}/S_2, where x_{OUTav} is the average value of all the x_{OUT} samples acquired within the intervals of 3 s. The results are included in Table 8.5. In the first row, the values $x_{OUT(av)}$ represent the average signal x_{OUT}. The second row presents the residual error ΔI_{DC}, while the voltage error ΔU_{DC} is shown in the third row. The voltage errors are considerably lower than 100 μV. Compared to the peak value of the phase voltages, the voltage errors in Table 8.5 do not exceed 0.125 parts per million.

8.4.5 Closed-Loop DC-Bias Suppression

High-precision dc-bias sensor can be used as the feedback device of an active dc-bias suppression system that includes a grid-side power converter as an actuator. Such a system is illustrated in Fig. 8.24 and discussed in detail in [9]. It is assumed that the ac grid introduces a dc-bias voltage U_{DCG}, which comes as a consequence of other grid resources. The dc-bias brought in from the grid causes the dc component U_{R0} of the voltage U_R across the sensing reactor. As a consequence, there is a dc component

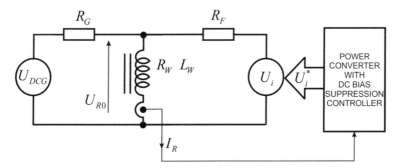

Fig. 8.24 Suppression of the bias by grid-connected power converter. The grid bias voltage U_{DCG} increases U_{R0} across the reactor. The power converter generates correction U_i which compensates the bias and makes sure that $U_{R0} = 0$

of the reactor current ($I_{mDC} = U_{R0}/R_W$) which produces the second harmonic of the reactor current.

Demodulator of Fig. 8.22 produces the signal x_{OUT}, which serves as the dc-bias feedback signal. It is assumed that the grid-side converter comprises the dc-bias suppression controller, which calculates the corrective dc voltage U_i, suited to act against the bias U_{R0} and adjusted so as to bring the reactor bias current I_{mDC} to zero. Corrective action of the grid-side inverter is performed by introducing some very small changes into the width of the PWM pulses, generated by the switching action of the power transistors.

The time constants L_G/R_G and L_F/R_F of Fig. 8.24 are lower than the time constant L_W/R_W of the sensing reactor by several orders of magnitude. They are also considerably lower than the sampling time $T_{SPL} = 20$ ms. Therefore, dynamic phenomena related to L_G and L_F are neglected and excluded from the subsequent analysis.

With $R_W \gg R_G$ and $R_W \gg R_F$, the dc voltage across the reactor is virtually unaffected by the reactor current:

$$U_{R0} \approx U_{DCG}\frac{R_F}{R_F + R_G} + U_i\frac{R_G}{R_F + R_G} \tag{8.45}$$

The parasitic voltage of the grid is denoted by U_{DCG}. It is changing rather slowly, and it remains almost constant within each sampling period T_{SPL}. The voltage U_i is the dc component of the output voltage supplied by the grid-connected power converter. This voltage is decided by the dc-bias controller (Fig. 8.25). A new U_i reference is calculated at each sampling instant nT_{SPL}, and this reference is maintained from nT_{SPL} until the next sampling instant $(n + 1)T_{SPL}$. Within the considered interval, the dc component of the reactor current $i_{R0}(t)$ changes in accordance to

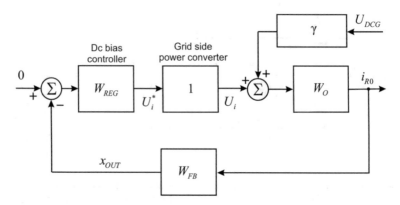

Fig. 8.25 Closed-loop dc-bias suppression that employs the sensing reactor as the feedback device and the grid-side inverter as an actuator

$$i_{R0}^{n+1} = i_{R0}^{n} \cdot e^{-\frac{R_W T_{SPL}}{L_W}} + \left(1 - e^{-\frac{R_W T_{SPL}}{L_W}}\right) \frac{U_{DCG}^{n} R_F + U_i^{n} R_G}{R_W(R_F + R_G)}. \tag{8.46}$$

Introducing the parameters α, β, and γ (8.47), the difference equation (8.46) assumes the form (8.48). The complex image $i_{R0}(z)$ is in (8.49), where $W_O(z) = \beta /(z - \alpha)$ is the pulse transfer $i_{R0}(z)/U_i(z)$.

$$\alpha = e^{-\frac{R_W T_{SPL}}{L_W}}, \beta = \frac{R_G(1 - \alpha)}{R_W(R_F + R_G)}, \gamma = \frac{R_F}{R_G}, \tag{8.47}$$

$$i_{R0}^{n+1} = i_{R0}^{n} \cdot \alpha + U_i^{n} \cdot \beta + U_{DCG}^{n} \cdot \beta \cdot \gamma. \tag{8.48}$$

$$i_{R0}(z) = \frac{\beta}{z - \alpha} U_i(z) + \frac{\beta}{z - \alpha} \cdot \gamma \cdot U_{DCG}(z). \tag{8.49}$$

The voltage U_i^{n} is calculated at instants $t = nT_{SPL}$, and it is applied on the interval $[nT_{SPL} .. (n + 1)T_{SPL}]$. The steady-state value of the signal x_{OUT} should be equal to zero. Thus, the reference value for x_{OUT} is zero. With the pulse transfer function of the controller $W_{REG}(z)$, the pulse transfer function of the voltage U_i is

$$U_i(z) = W_{REG}(z)[0 - x_{OUT}(z)]. \tag{8.50}$$

The signal x_{OUT} is proportional to the dc-bias current of the reactor i_{R0}, $x_{OUT} \approx S_2 i_{R0}$. The sample x_{OUT}^{n} is calculated at instants $t = nT_{SPL}$ as an average value of the signal x_{IN} over the past sampling interval, and it depends on the average value of $i_{R0}(t)$ on the same interval. The average value of i_{R0} on the interval $[nT_{SPL} .. (n + 1) T_{SPL}]$ can be approximated by $(i_{R0}^{n+1} + i_{R0}^{n})/2$. In z domain,

$$x_{OUT}(z) = S_2 \frac{z + 1}{2z} i_{R0}(z) = W_{FB}(z)i_{R0}(z). \tag{8.51}$$

The block diagram of the closed-loop dc-bias suppression system is summarized in Fig. 8.25. The steady-state value of the dc voltage across the sensing reactor should remain zero in the presence of the dc-bias U_{DCG}. This requires an integrator within the control loop. Considering the PI controller with the gains K_P and K_I, and introducing the relative gains $p = K_P \cdot \beta \cdot S_2/2$ and $i = K_I \cdot \beta \cdot S_2/2$, the closed-loop transfer function $W_{SS}(z) = i_{R0}(z)/x^*_{OUT}(z)$ is

$$W_{SS}(z) = \frac{2}{S_2} \frac{(p+i)z^2 - pz}{z^3 + z^2(-1 - \alpha + p + i) + z(\alpha + i) - p}. \qquad (8.52)$$

The step response of the dc voltage U_{R0} across the reactor is defined by the function $W_{DCG}(z) = U_{R0}(z)/U_{DCG}(z)$:

$$W_{DCG}(z) = \frac{\gamma \cdot z \cdot (z-1) \cdot (z - \alpha) \cdot R_G/(R_G + R_F)}{z^3 + z^2(-1 - \alpha + p + i) + z(\alpha + i) - p}. \qquad (8.53)$$

The closed-loop performance depends on the closed-loop poles, which are the roots of the characteristic polynomial $f(z) = z^3 + z^2(-1 - \alpha + p + i) + z(\alpha + i) - p$. The three poles are decided by the gains p and i. Design goal of the gain setting procedure is to obtain a fast, well-damped response with a short settling time [9]. At the same time, it is necessary to obtain sufficient robustness. The equivalent resistance of the grid (R_G) is variable, and it has significant impact on the closed-loop response. Therefore, it is necessary to provide sufficient stability margin and to maintain the response character in the presence of R_G changes. This request can be met by prescribing the minimum vector margin [9] of $VM_{min} = 2/3$. The optimum gains are found by searching the p-i plane for the pair (p, i) which provides short settling time and $VM > VM_{min}$. The gains that meet both requirements are $p_{opt} = 0.18$, $i_{opt} = 0.026$. With optimum gains and with $T_{SPL} = 20$ ms, the closed-loop transfer function drops to -3 dB at the bandwidth frequency of $f_{BW} = 6.65$ Hz [9]. The vector margin is $VM = 0.68$, thus accommodating the grid resistance changes by more than 3 times. The step response overshoot is 27%, and the rise time is 60 ms.

The experimental setup of the active dc-bias suppression system is detailed in [9]. The block diagram of the setup is given in Fig. 8.26. The three-phase mains are connected through an isolation transformer which prevents the impact of the unknown parasitic dc-bias from the actual grid. A variable-ratio three-phase transformer is used to change the ac voltage level. Well-known and controlled dc-bias is introduced by closing the switch SW1. Two very large resistances (R_{11} and R_{12}) are used to inject small dc currents, obtained from the dc-bus of the grid-side power converter. When the dc-bias is injected, the sensing reactors detect the parasitic dc current from the corresponding even harmonic and its phase. The closed-loop controller introduces the corrections by affecting the PWM pulses of the grid-side converter and providing small dc voltage components between the output terminals of the inverter. These corrections are suited to suppress the dc-bias within the core

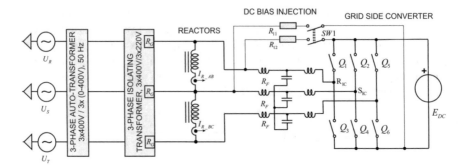

Fig. 8.26 Dc-bias detection and suppression in three-wire three-phase system using two sensing reactors and two dc-bias controllers. The dc-bias is introduced by closing the switch SW1

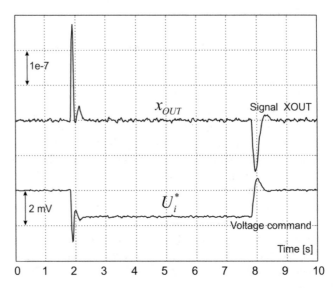

Fig. 8.27 Experimental traces of the signal x_{OUT} and the voltage command U_i^* obtained from the system in Fig. 8.26. The dc-bias is introduced by closing the switch $SW1$

and to bring the core into zero-bias condition. In turn, that will suppress the dc-bias between the grid-connection points.

Dynamic response is tested by closing and opening the switch SW1. Experimental traces of the signal x_{OUT} and the voltage command U_i^* obtained from the system in Fig. 8.26 are shown in Fig. 8.27. After 2 s, the dc-bias is introduced by closing the switch SW1. The bias is removed after 8 s. The transient response has the settling time in agreement with the pulse transfer function of (8.53), while the waveforms comprise the noise and include the effects of the system nonlinearity, observed from the differences between the two transients.

Due to imperfections, the circuit of Fig. 8.26 retains a small residual error. The error of the dc-bias suppression system is tested in steady-state condition, with SW1

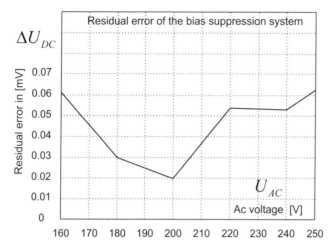

Fig. 8.28 Residual error of dc-bias suppression system

opened and with no bias. The error is determined from the voltage U_i as $U_{R0} = U_i \cdot R_G/(R_G + R_F)$. The measurement of U_i is accomplished with appropriate filtering and averaging. The results are plotted in Fig. 8.28. Residual errors in Fig. 8.28 are lower than 65 µV. They are obtained with the sensing reactor core of $m_{Fe} = 0.304$ kg. Competitive solutions [8] have residual errors lower than 100 µV, obtained with the reactor core of $m_{Fe} = 0.408$ kg. Thus, the introduction of the optimum core form improves the precision-to-weight ratio by 1.93 times. In closed-loop operation, the dc-bias voltage across the grid-connection terminals is suppressed below 65 µV, the level acceptable even with the most recent distribution transformers.

Question (8.1) For the phase-locked loop of Fig. 8.6, with the closed-loop transfer function of (8.6), determine the gains to obtain two real closed-loop poles at 2 Hz.

Answer (8.1) The closed-loop transfer function of the phase-locked loop is

$$W_{CL}(s) = \frac{\varphi_{PLL}(s)}{\varphi_L(s)} = \frac{s \cdot k_p k_{PD} k_{VCO} + k_i k_{PD} k_{VCO}}{s^2 + s \cdot k_p k_{PD} k_{VCO} + k_i k_{PD} k_{VCO}}.$$

In order to have two real poles $p_1 = p_2 = \omega_p = 4\pi$ [rad/s], the value of $k_i k_{PD} k_{VCO}$ has to be equal to ω_p^2, while the value of $k_p k_{PD} k_{VCO}$ has to be equal to $2\omega_p$.

Question (8.2) A large two-pole synchronous generator runs at no load, with constant excitation current. With the power angle of $\pi/2$, the generator will deliver the maximum power of 300 MW. The oscillations caused by the load step have the frequency of 1.5 Hz. It is necessary to estimate the equivalent inertia J of the generator.

Answer (8.2) From Sect. 8.2.1 and Figs. 8.7 and 8.8, the electromagnetic torque of the generator (i.e., the braking torque) can be approximated by $T_G = T_{MAX} \sin(\delta)$, where δ is the power angle (angular advance of the rotor electromotive force with respect to the stator voltage) while T_{MAX} is the torque obtained with $\delta = \pi/2$. If the generator runs at no load, the power angle is close to zero. Thus, the torque T_G can be approximated by $T_G = T_{MAX} \cdot \delta$. The maximum torque is obtained from the maximum power and rated speed, $T_{MAX} = 300[MW]/314.15[rad/s] = 954,000$ Nm.

Neglecting any other torque components at no load condition, the rotor speed changes depend only on T_G. At the same time, $\delta = \theta_m - \varphi_L$, where θ_m is the rotor angle while φ_L is the phase of the grid voltages. Thus,

$$J\frac{d\omega_m}{dt} = J\frac{d^2\theta_m}{dt^2} = J\frac{d^2\delta}{dt^2} + J\frac{d^2\varphi_L}{dt^2} = -T_{MAX}\delta.$$

Assuming that the line frequency is constant, the second derivative of φ_L is equal to zero. By considering the previous differential equation and introducing the Laplace transform, one obtains

$$J\frac{d^2\delta}{dt^2} + T_{MAX}\delta = 0 \rightarrow Js^2\delta(s) + T_{MAX}\delta(s) = 0.$$

The roots of the equation

$$f(s) = J \cdot s^2 + T_{MAX} = 0$$

are given in

$$p_{1/2} = \pm j \cdot \sqrt{\frac{T_{MAX}}{J}}.$$

With $|p_{1/2}| = 2\pi \cdot 1.5 = 3\pi$, $J = T_{MAX}/(3\pi)^2 = 10,740$ kgm^2.

Question (8.3) The wind power station has a wind turbine, electrical generator, generator-side converter, dc-bus with the voltage E_{DC} and bus capacitor C_{DC}, and grid-side inverter which takes the power P_{DC} from the dc bus and injects the power $P_{AC} \approx P_{DC}$ into ac grid. The system is depicted in Fig. 8.13. The power $E_{DC} \cdot i_{GE}$ that comes from the generator depends on the wind speed, and it cannot be changed at will. The power $P_{DC} = E_{DC} \cdot i_{GR}$ is determined by the PI controller suited to maintain the bus voltage E_{DC} constant. Determine the gains of the controller so as to obtain the response characterized by the closed-loop poles with natural frequency ω_n and with damping factor of ξ.

Answer (8.3) The dc-bus voltage is increased by i_{GE} and decreased by i_{GR}:

$$C_{DC}\frac{dE_{DC}}{dt} = i_{GE} - i_{GR} \rightarrow E_{DC}(s) = \frac{1}{sC_{DC}}[i_{GE}(s) - i_{GR}(s)].$$

The current i_{GE} is determined by the wind speed, and it cannot be changed at will. From the point of view of the dc-bus control, it is considered an external disturbance.

The current i_{GR} is equal to $P_{DC}/E_{DC} = P_{AC}/E_{DC}$, and it is determined by the ac currents injected into the grid from the grid-side power converter. Thus, the digital current controller has the ability to adjust the ac currents so as to provide the desired current i_{GR}. At this point, it is necessary to obtain the reference value for i_{GR}. The problem statement considers the PI controller as the means of discriminating the errors in the dc-bus voltage ($\Delta E_{DC} = E^* - E_{DC}$) and calculates the desired current i_{GR}:

$$i_{GR}^* \equiv i_{GR} = \left(K_P + \frac{K_I}{s} \right) \Delta E_{DC}.$$

In the above expression, it is assumed that the controls guarantee that the actual current i_{GR} gets equal to the reference. The characteristic polynomial of the system is obtained from

$$1 + \frac{1}{sC_{DC}} \left(K_P + \frac{K_I}{s} \right) = 0, \rightarrow s^2 + s\frac{K_P}{C_{DC}} + \frac{K_I}{C_{DC}} = 0.$$

The gains can be obtained from $\omega_n^2 = K_I/C_{DC}$ and $2\xi\omega_n = K_P/C_{DC}$.

References

1. Picher P, Bolduc L, Dutil A, Pham VQ (1997) Study of the acceptable DC current limit in core-form power transformers. IEEE Trans Power Deliv 12(1):257–265
2. HV/LV distribution transformers, TRIHAL cast resin dry type transformers 160 to 2500 kVA (2005) France Transfo, Schneider Electric Industries SAS. http://mt.schneider-electric.be/
3. Cast resin dry type distribution transformers (2014) Lemi-Trafo Tranformers. www.lemi-trafo.com
4. Amorphous alloy transformers: underground amorphous alloy transformer (2011) China Power Equipment Inc. http://www.cet.sgcc.com.cn/
5. Mousavi SA, Engdahl G, Agheb E (2011) Investigation of GIC effects on core losses in single phase power transformers. Arch Elect Eng 60(1):35–47
6. Buticchi G, Lorenzani E, Franceschini G (2011) A DC offset current compensation strategy in transformerless grid-connected power converters. IEEE Trans Power Deliv 26(4):2743–2751
7. Buticchi G, Lorenzani E (2013) Detection method of the dc bias in distribution power transformers. IEEE Trans Ind Electron 60(8):3539–3549
8. Vukosavic SN, Peric LS (2015) High-precision sensing of DC bias in AC grids. IEEE Trans Power Deliv 30(3):1179–1186
9. Vukosavic SN, Peric LS (2017) High-precision active suppression of DC bias in AC grids by grid-connected power converters. IEEE Trans Ind Appl 64(1):857–865

Bibliography

1. Kimbark EW (1956) Power system stability, vol. III, Synchronous machines. Wiley, New York
2. White DC, Woodson HH (1959) Electromechanical energy conversion. Wiley, New York
3. Vukosavić SN (2007) Digital control of electrical drives. Springer, New York
4. Wildi T (2006) Electrical machines, drives and power systems. Pearson Prentice Hall, Upper Saddle River
5. Vukosavić SN (2013) Electrical Machines. Springer, New York, USA. ISBN 978 1-4614-0399-9, Library of Congres 2012944981

© Springer International Publishing AG, part of Springer Nature 2018   259
S. N. Vukosavic, *Grid-Side Converters Control and Design*, Power Electronics and Power Systems, https://doi.org/10.1007/978-3-319-73278-7

Index

© Springer International Publishing AG, part of Springer Nature 2018   261
S. N. Vukosavic, *Grid-Side Converters Control and Design*, Power Electronics
and Power Systems, https://doi.org/10.1007/978-3-319-73278-7

Printed in the United States
By Bookmasters